Progress in Optical Science and Photonics

Volume 23

The purpose of the series Progress in Optical Science and Photonics is to provide a forum to disseminate the latest research findings in various areas of Optics and its applications. The intended audience are physicists, electrical and electronic engineers, applied mathematicians, biomedical engineers, and advanced graduate students.

Arijit Saha · Arindam Biswas · Kankat Ghosh ·
Nilanjan Mukhopadhyay
Editors

Optical to Terahertz Engineering

Springer

Editors
Arijit Saha
Department of Electronics
and Communication Engineering
B. P. Poddar Institute of Management
and Technology
Kolkata, India

Kankat Ghosh
Department of Electrical Engineering
Indian Institute of Technology Jammu
Jammu, Jammu and Kashmir, India

Arindam Biswas
School of Mines and Metallurgy
Kazi Nazrul University
Asansol, West Bengal, India

Nilanjan Mukhopadhyay
Department of Electronics
and Communication Engineering
Global Institute of Management
and Technology
Nadia, India

ISSN 2363-5096 ISSN 2363-510X (electronic)
Progress in Optical Science and Photonics
ISBN 978-981-99-0227-9 ISBN 978-981-99-0228-6 (eBook)
https://doi.org/10.1007/978-981-99-0228-6

This Springer imprint is published by the registered company Springer Nature Singapore Pte Ltd.
The registered company address is: 152 Beach Road, #21-01/04 Gateway East, Singapore 189721, Singapore

Preface

With the advancements in areas like optical communication, signal processing, imaging systems, etc. researchers, industry persons and end users are required to be familiarized with the progress in these sectors. In recent years, the almost unexplored domain of the THz range of electromagnetic spectrum has paved the way for terahertz technology. A wide range of applications using THz techniques such as terahertz time-domain spectroscopy (THz-TDS), biological, medical and pharmaceutical sciences, explosives inspection, Information and Communication Technology (ICT) sector and many more have the potential to be the technology of future. The present title "Optical to Terahertz Engineering" focuses on the recent developments in Optical and THz technology with an emphasis on their basic fundamentals. It illuminates on different materials required for THz and optical engineering, explains methodologies and challenges of imaging and secured communication in both THz and optical domains, and also discusses the applications of fiber optics in different domains. This book serves as a guide to THz and optical technology for new researchers in various fields. Many different disciplines, such as phase controllers for THz-Spectro-Polarimeters and photo-emission technology from heavily doped THz materials, involve the recent developments of THz technology. On the other hand, advancements and future prospects of Lithium Niobate-Based polarization phase modulator, photonic crystal fiber, Optical Code Multiple Access System, satellite image correction and thermal image processing emphasizes optical technologies. The focus of the current title mainly lies in the practical applications of THZ and optical engineering. Optical to Terahertz Engineering is an ideal book for a very vast audience from basic science to engineering and technology experts and learners.

This book has the potential to become a textbook for engineering and Post Graduate programs in science, and also for researchers. This title also serves the common public interest by presenting new methods of integration into society.

Kolkata, India Arijit Saha
Asansol, India Arindam Biswas
Jammu, Jammu and Kashmir, India Kankat Ghosh
Krishnagar, India Nilanjan Mukhopadhyay

Contents

About the Editors

Prof. Arijit Saha Professor, Department of Electronics & Communication Engineering, B.P. Poddar Institute of Management & Technology, Kolkata, West Bengal, 700052, India.

Short Biography: Professor Saha received his B.Sc. (Physics) from the University of Calcutta in 1996. He received his B.Tech. in 1999 and M.Tech. in 2004, both in Optoelectronics from the University College of Technology, Calcutta University. He received his Ph.D. (Technology) from the same niversity in January 2013. He has more than twenty-three years of experience in teaching and industry. Currently, he is working as a Professor in the Department of Electronics & Communication Engineering at B.P. Poddar Institute of Management and Technology, Kolkata. He is also attached as a Visiting Professor with the University of Calcutta. He has 44 research publications of both international and national repute. He has authored 4 books published by famous publishing houses like Infinity Science Press, USA; University Science Press, Delhi; LAP Germany; and Pearson Education. He has also authored 8 book chapters published by reputed publishers like Elsevier and Springer. Professor Saha has two patents published under his credit. He has also organized and chaired different International and National Conferences. His research interest includes optical communication, birefringent network, optical and THz signal and image processing, machine learning and renewable energy. Professor Saha acted as a reviewer for several reputed journals. Professor Saha is a Fellow of The Institution of Electronics and Telecommunication Engineers

(IETE), and Optical Society of India (OSI). He is also the life member of the Indian Society for Technical Education (ISTE) and International Association of Engineers (IAENG). e-mail: arijitsah@gmail.com

Dr. Arindam Biswas Associate Professor, School of Mines and Metallurgy, Kazi Nazrul University–713305, India.

Short Biography: Dr. Arindam Biswas received his M.Tech. degree in Radio Physics and Electronics from University of Calcutta, India, in 2010 and his Ph.D. from NIT Durgapur in 2013. He has worked as a Postdoctoral Researcher at Pusan National University, South Korea, with a prestigious BK21PLUS Fellowship, Republic of Korea 2015, DST-JSPS Invitation Research Grant Award in 2020, and DST-ASIAN Invitation Research Grant 2021 for IoT-based Environmental Monitoring System in Underground Mines. He has been a Visiting Professor at Research Institute of Electronics, Shizuoka University, Japan. He has been selected for IE(I) Young Engineer Award 2019–20 from Institute of Engineers (I), India. Dr. Biswas has 12 years of experience in teaching research and administration. Presently, Dr. Biswas is working as an Associate Professor in School of Mines and Metallurgy at Kazi Nazrul University, Asansol, WB, India. He has 53 technical papers in different journals and 35 conference proceedings and 7 books, 11 edited volumes and 12 book chapters with international repute. Dr. Biswas received a research grant from Science and Engineering Research Board, Govt of India, under Early Career Research Scheme for research in Terahertz-based GaN Source. He has also received Research Grant from Centre of Biomedical Engineering, Tokoyo Medical and Dental University in association with RIE, Shizuoka University, Japan, for the study of biomedical THz Imaging based on WBG semiconductor IMPATT Source for two consecutive years 2019–20 and 2020–2021. Dr. Biswas is also associated with KNU-UGC Stride Grant for developing an IoT-based system for Mines Safety. Presently, Dr. Biswas is serving as an Associate Editor of Cluster Computing, Springer (SCI Indexed). Dr. Biswas has produced 6 Ph.D. students in different topics of applied optics and high-frequency semiconductor devices. He has organized and chaired different International Conferences in India and abroad. His research interest is in carrier transport in low dimensional systems and electronic devices, nonlinear optical communication and THz Semiconductor Source. Dr. Biswas acted as a reviewer for reputed journals, a member of the Institute of Engineers (India) and a Regular Fellow of Optical Society of India. e-mail: mailarindambiswas@yahoo.co.in

Dr. Kankat Ghosh Assistant Professor, Department of Electrical Engineering, Indian Institute of Technology Jammu, Jammu, Jammu and Kashmir 181221, India.

Short Biography: Kankat Ghosh received his B.Sc. degree in Physics from Calcutta University, India, in 2006. After that, he received his B.Tech. and M.Tech. degrees in Radio Physics and Electronics from University of Calcutta, India, in 2009 and 2011, respectively. He received his Ph.D. from Indian Institute of Technology Bombay in June 2018. He has around 10 years of experience in research. At present, he is working as an Assistant Professor in the Department of Electrical Engineering, Indian Institute of Technology Jammu in India. He is a life member of the Materials Research Society of India (MRSI) and a member of IEEE. His research interest is the epitaxial growth and fabrication of optoelectronic and electronic devices. He has authored/co-authored 13 research articles in peer-reviewed international (SCI) journals and contributed to (presentations/articles) around 20 international conferences. e-mail: kankat.ghosh@iitjammu.ac.in

Dr. Nilanjan Mukhopadhyay Assistant Professor, Department of Electronics and Communication Engineering, Global Institute of Management and Technology–741102, India.

Short Biography: Nilanjan Mukhopadhyay received his B.Sc. degree in Physics from Calcutta University, India, in 2006. After that, he received his B.Tech. and M.Tech. degrees in Optics & Optoelectronics from Calcutta University in 2009 and 2011, respectively. He has more than 10 years of experience in teaching and industry. At present, he is working as an Assistant Professor in the Department of Electronics and Communication Engineering, Global Institute of Management and Technology, India. He is also pursuing his Ph.D. in the Department of Applied Optics and Photonics, University of Calcutta, India, and has submitted a thesis. His current research field includes birefringent networks, polarization optics, THz system design, satellite image processing, optical communication and integrated optics. He has about 20 research publications in peer-reviewed journals and at national and international conferences. Mr. Mukhopadhyay has one patent published under his credit. He has served as a reviewer in SCOPUS indexed journals and different International and National Conferences. He acted as an active member of the organizing committee of several national and international conferences and workshops. He is a life member of the Optical Society of India (OSI). e-mail: nilu.opt@gmail.com

Introduction to Terahertz Imaging Applications

Semanti Chakraborty and Kanik Palodhi

1 Introduction

Terahertz frequency domain is one of the most anticipated but largely unexplored regions of electromagnetic waves. As the name suggests, the frequency domain is generally considered to be between 0.1 THz and 10 THz, where 1 THz $= 10^{12}$ Hz corresponding to a wavelength range varying between 0.3 mm and 30 μm. Essentially, this domain is within high-frequency electronics close to the microwave region (~10 GHz/0.3 mm) and far-infrared optics ~100 THz/0.3 μm as shown in Fig. 1 [1–9]. Terahertz, therefore, closes the gap between these two mature technologies and provides the best of both worlds, so to speak, by making itself available to both communication and imaging [10–17].

One of the main reasons for the terahertz domain being unexplored is the unavailability of inexpensive generators or sources in this region. Terahertz domain corresponds to a photon energy of 4.1 meV which belongs to the energy range of thermal excitation regions of the semiconductor [18–22]. The traditional lasing option used in optics or novel use of different diodes in microwave regions owing to their bandgap cannot be used here as it is. This led to a technology gap known as the *terahertz gap*. In fact, before the coinage of the term terahertz/THz, it was often referred to as far-optics clustering it with far-infrared optics [23–25].

Despite the lack of suitable sources, some of the advantages of terahertz technologies are evident if interpolated from the proven principles (or technologies) in optics and microwave. One of the most advantageous aspects of this technology is that the terahertz probe beams are non-ionising and invisible to the eye [2, 22, 23]. Secondly, in this domain, using coherent detection techniques, the exact amplitude

S. Chakraborty
Department of Electronics and Communication Engineering, Amity University, Kolkata, India

K. Palodhi (✉)
Department of Applied Optics and Photonics, University of Calcutta, Kolkata, India
e-mail: kpaop@caluniv.ac.in

A. Saha et al. (eds.), *Optical to Terahertz Engineering*, Progress in Optical Science and Photonics 23, https://doi.org/10.1007/978-981-99-0228-6_1

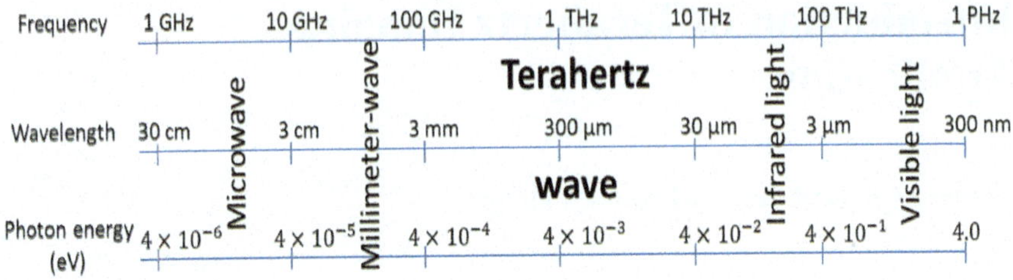

Fig. 1 Position of the terahertz wave in electromagnetic spectrum

and phase of the electric field can be determined leading to important applications in imaging, instrumentation and metrology [26–34]. During imaging, this technology also has the power to penetrate through layers of different materials including clothes [35–40]. Recently, body scanners developed using terahertz have been much talked about. It is a critical application in terms of security checking since prohibited items can be tracked, however, it is also ethically questionable as it leaves a trail of the human body together with its belongings. Apart from this, the interaction of terahertz waves (T waves) with matter can help investigate chemical compounds by applying spectroscopic techniques, which in turn, be of use in biomedical and pharmaceutical applications [41–55].

2 Instrumentation

In recent years, due to the advancement of technology, thanks to consistent efforts from both academia and industry, this domain is beginning to claim its rightful place closing the so-called 'terahertz gap'. The most important aspect of THz technology that is propelling life-changing applications is the instrumentation containing different sources, detectors and other passive components. These are developed from emulating matured photonic and microwave technologies.

2.1 Sources

Currently, the available THz sources can be broadly classified into two groups depending upon their emission mode and operating frequency—continuous (CW) and pulsed (time-domain/TD) [8, 18–21]. Each category has different advantages and disadvantages prompting different usage. The CW sources developed using technologies pertaining to millimetre waves essentially scale up the frequencies and can reach up to the operating frequency of 5 THz in the case of backward-wave oscillators [22, 56]. A different approach can be used by applying the down-conversion principle commonly used in photonic waves. In this case, two sources differing in frequency can also be used to generate T-waves [22, 57]. Typically, CW sources are narrowband providing greater power compared to pulsed sources. They are more sensitive and are suitable for heterodyning applied to the communication of non-destructive testing applications.

The pulsed/TD sources, on the other hand, are predominantly obtained from the conversion of optical signals to T-waves. By detecting the electromagnetic transient generated from a picosecond, laser pulse T-waves are generated. Since the short pulse is composed of multiple frequencies, it is typically applied to investigate spectral characteristics or ultrafast phenomena [57–62].

2.2 *Detectors*

THz detectors are usually categorised into three groups—thermionic, photoconductive antenna and electro-optic. Recent developments have made sure that accurate measurements of amplitude and phases can be done successfully using these techniques, particularly, photoconductive and electro-optic. Most of the commercially available THz systems, though only a few, have these types of detectors used [22, 23, 63–67].

2.3 *Optics*

Another important aspect of THz systems is the beam formation optics which is essential for higher throughput and diffraction-limited spot generation. This is challenging because (a) terahertz wavelengths cannot be treated as negligible compared to the size of the optical elements and (b) optical elements need to be achromatic over the entire THz region for flat phase response. This is generally achieved by using dispersionless substrate lenses made of highly resistive ($10^4\ \Omega$) silicon (refractive index $n = 3.42$) attached to the transmitter and receiver. Typically, the preferred design employs an aplanatic hyper hemispherical lens (a variant of hemispherical lens) which reduces aberrations such as astigmatism and increases effective aperture providing smooth collimation of the T-rays [2, 22, 68–73].

Another very important optical component is a parabolic mirror which is commonly used for beam traversal and focussing. This is typically used with a metal coating and has stringent specifications to be met [2]. Apart from beam coupling, recent developments have been made for beam engineering using traditional concepts of diffraction gratings, diffractive lenses, Fresnel lenses and similar components using spatial terahertz modulators or holograms [74–76]. These components can be made relatively cost-effectively using graphite, 3D-printed passive beam guides, high-density polyethylene (HDPE) and high-resistivity silicon [77–81]. In recent years, meta-materials have been developed for manipulating the beam using meta-surface amplitude, phase and polarisation of the T-beam [82–85]. Some of the interesting applications of the use of metasurfaces are (a) vortex-beam generation [84], (b) novel components using polarisation conversion [86] and (c) polarisation-controlled superfocusing [87].

2.4 Signal Acquisition and Important Parameters

The signal acquisition process, here, is one of the most important aspects of the instrumentation. Traditionally, the delay in the generated pulse is scanned over the detecting pulse to measure the average photocurrent as a function of the delay. Essentially, this is a process of convolution of the temporal shape of the pulse with the generated waveform and this was previously achieved by a lock-in amplifier with a chopper wheel [23]. This process was not fast enough and was replaced by a scanning optical delay line as suggested by Hu and Nuss in their seminal paper [10]. The most common parameters employed to measure the performance of signal acquisition systems are dynamic range and signal-to-noise ratio (SNR). DR is defined as the difference between the maximum to minimum signal, whereas SNR provides a measure of the minimum detectable signal and sensitivity of the device.

Current measurement methods can have an SNR of 1000 or more for imaging applications. In the case of spectral imaging, a raster scan or a scanning mirror is used for a certain range of an area of the object under investigation. DR and SNR can be evaluated from the amplitude spectrum upon Fourier transform if spectroscopic data are analysed. Another important parameter for the spectroscopic system is spectral resolution given by twice the ratio of the velocity of light c and effective delay [2, 22, 88, 89]. The process of signal acquisition is automated by software, and even accessible spectral databases have been developed for comparison purposes.

2.5 Signal Processing

After the acquisition of the signal, image processing software is employed for the extraction of the amplitude and phase information. Different transform-based approaches such as Fourier, Hilbert or other approaches such as synthetic aperture imaging, reflectometry, etc. are quite popular for this purpose. Typical image processing operations, such as edge detection, histogram processing, etc. can also be performed. Recently, machine learning-based image processing techniques have also been successfully employed for different information [90].

3 Imaging

The first instance of THz images was reported by Hartwick et al. in 1976 using an optically pumped THz laser [91]. Then, in their seminal paper, Hu and Nuss presented the idea of THz imaging using a femtosecond laser in 1995 [10]. As described in the introduction, THz waves offer unique capabilities of penetrating through non-polar and non-metallic objects such as paper, plastics, ceramics, clothes, etc. which makes it suitable for imaging and spectroscopy used in industrial imaging such as non-destructive evaluation (NDE) [7, 92], material characterisation [93, 94], medical diagnostics [9, 12], cultural heritage objects such as paintings, etc. [95, 96]. Absorption of THz waves in water is very high which is of great importance for imaging of organic materials or medical imaging [97–99]. Apart from metal and water, most other materials have a unique signature in the spectral domain (~THz). This can be combined with the THz imaging setup for the detection of special reagents such as explosives, medicines, etc. [14, 100].

In Fig. 2, it can be seen that three objects, (a) a smart card, (b) a coleus leaf and (c) a razor, partially covered by a polythene sheet have been imaged using THz imagers. Clearly, the THz waves have penetrated non-metallic materials and have been reflected from the metals. For organic samples, the water molecules have absorbed the T waves, therefore, it appears dark in the regions where water is present.

3.1 *Imaging with Continuous-Wave (CW) THz Waves*

Many of the initial systems in THz domain were based on the principle of CW operations. Typically, photomixers, THz parametric oscillators (TPO), backward wave oscillators, QCLs, etc. are used as sources with a frequency of operation close to 0.5 THz, and heterodyne detectors are used as receivers. The instrumentation of

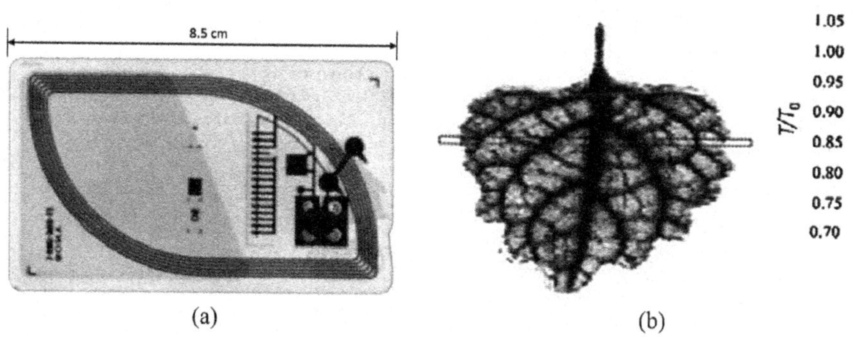

Fig. 2 Examples of THz images—**a** THz transmission image of a contactless smart card embedded with a metallic circuit [101]; **b** THz image of a Coleus leaf where the grayscale is correlated with water content, with darker shade indicating more water [69]

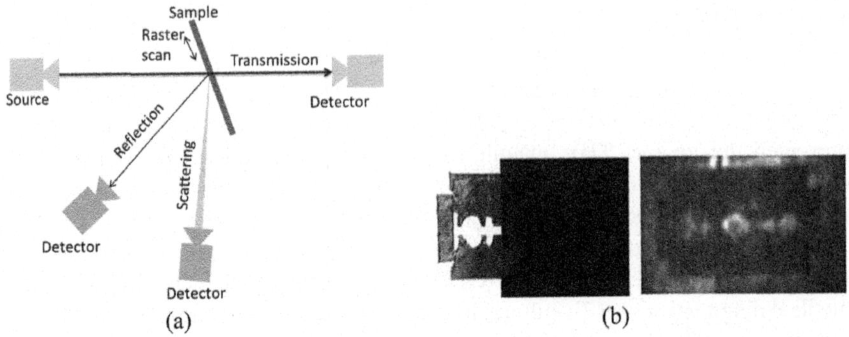

Fig. 3 **a** Scheme of CW operations using raster scan; **b** Scaled razor blade covered by a black polyethylene sheet—a photograph and a THz image using 2.52 THz, 10 mW radiation from a gas laser with a microbolometer detector. In the THz image, the covered part of the blade is clearly visible [102]

the CW imaging systems is very critical to the specific needs of the application. For example, for higher sensitivity, cryogenic detectors could be required or for moving objects, high frame rates of the detectors may be necessary. There are numerous applications such as body scanners, non-destructive testing units, etc. which have been developed based on this principle.

Raster scan, as shown in Fig. 3a, is the most commonly used scheme for CW operations [2, 103]. Here, the THz wave from the source passes through the sample, and depending upon the application, the reflected/scattered or the transmitted beams are recorded. An example of a typical security system is shown in Fig. 3b, where the two images of a razor blade, partially covered by a black polythene sheet, are shown. The first one, the photograph, could not reveal the covered part, whereas in case of THz imaging, the covered parts are visible.

3.2 *Imaging with THz Pulses*

CW operation, as described in the previous section, has to be designed for specific purposes and lacks flexibility. The THz pulsed system, essentially, THz time-domain spectroscopy can be designed to be flexible, compact and cost-effective. In addition, by measuring - (a) the diminished amplitude of the sub-picosecond pulses after passing through the object and (b) the phase change of the pulse, crucial information can be extracted about the samples [2, 22]. In fact, recent commercial instruments from Teraview or Picometrix Inc. have already developed these techniques [104].

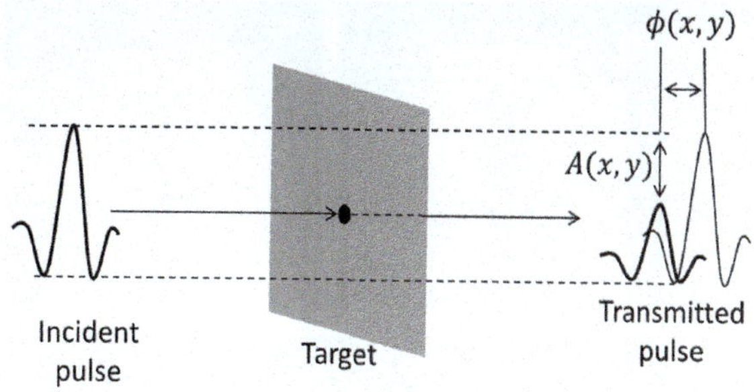

Fig. 4 Amplitude and phase imaging scheme

3.2.1 Amplitude and Phase Imaging

The basic scheme of the THz pulsed mode imaging (THz-TDS) is described here where both amplitude and phase imaging can be performed. Initially, the THz pulse is focussed onto the sample, and the change in amplitude $A(x, y)\phi(x, y)$ of the transmitted (or reflected, depending upon the geometry) pulse is measured as shown in Fig. 4. These changes will depend on the material properties of the sample such as refractive index $n(x, y)$, absorption coefficient $\alpha(x, y)$ and thickness $d(x, y)$. For the investigation of the entire target region, a raster scan is applied.

It is evident that the THz-TDS offers much more information than the CW systems because from each image-pixel, changes in a pulse can be monitored to know about the properties of the sample. Additionally, the size of the focal spot on the object is inversely proportional to the frequency domain [mitt]. The basic idea, here, is to find out the variation in peak-to-peak amplitude in the time domain and track the change in phase. For spectral information, typically, Fourier transform-based methods are applied to the pulses integrated over a desired frequency band. Apart from this, many different transforms such as wavelet and Hilbert have been applied to extract amplitude and phase information [2, 105].

3.3 Comparison

CW imaging operation is a well-established method and proved to be better when it comes to long-range imaging. This is due to the fact that at a particular frequency band between 1.0 THz and 5.0 THz, T-beam is nearly transparent in the atmosphere. Combining the QCL source and microbolometer camera, good quality CW images can be obtained even at a distance of 25 m [106]. For enhanced spatial resolution, however, objects need to be put at a closer distance (see Fig. 5).

The THz-pulsed operation as described, however, provides better imaging in terms of accuracy and versatility with a need for complex data processing capability. It definitely provides much more information about the objects under investigation. Karpowicz et al. actually compared the two techniques by investigating defects in

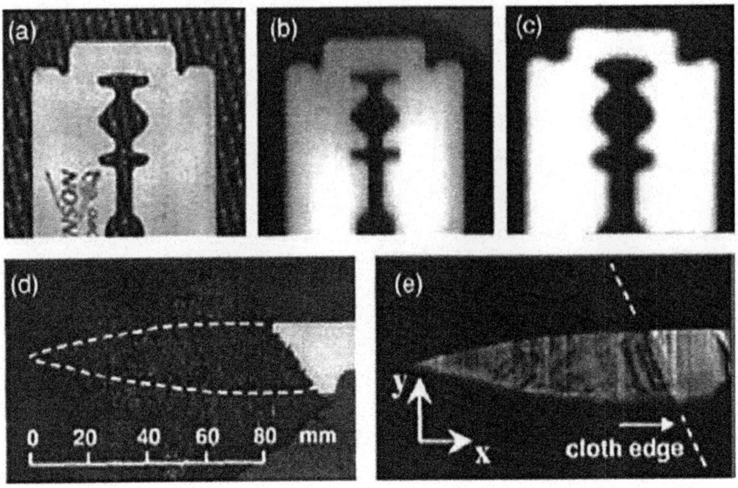

Fig. 5 Comparison of CW and TDS—Images of a razor blade, **a** photograph, **b** pulsed THz image, **c** CW THz image at 0.53 THz; Images of a ceramic knife, concealed underneath a denim cloth, **d** photograph and **e** CW THz image at 0.53 THz [107]

foam used in space shuttle keeping similar beam sizes for both methods [108]. They have concurred that the simplicity of CW operation could be preferred if details of the samples are required. One definite advantage of the pulsed TDS technique is that it could further be extended to material characterisation, tomography, surface topography measurements and a variety of other applications.

4 THz Tomography

In case of a multi-layered sample, the sub-picosecond pulses get reflected from the interfaces of these layers as shown in Fig. 6a. By detecting these pulses, depth-resolved imaging of the entire sample can be made by bringing the unexplored third dimension to the fore [109–115]. Since THz waves can penetrate through most of the non-polar and non-metallic objects, this technique proves to be highly beneficial for different industrial applications. Assuming a delay of Δt of the reflected pulse (for normal incidence) from the layer of thickness d and refractive index n, the thickness can be expressed as

$$d = c\Delta t/2n \tag{1}$$

where c is the speed of light.

The typical depth resolution obtained has reached submicron levels now [115]. As an example, a tomographic image of a floppy disc is shown in Fig. 6b. The technique

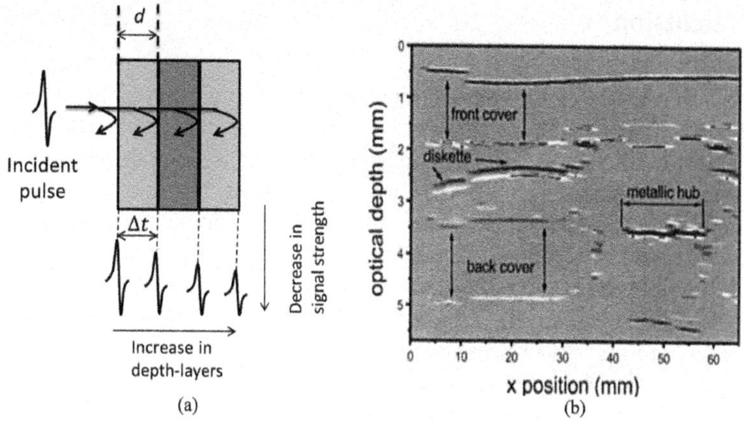

Fig. 6 **a** THz tomography scheme and **b** example of a THz depth image of a floppy disc at total reflected power after post-processing [116]

has been already applied to various industries for measuring coatings on tablet and for the inspection of paints, among many other applications. In many cases, optical coherence tomography cannot be used due to low penetration.

5 Latest Developments

Recent developments in terahertz imaging are essentially trying to incorporate the concepts from photonics and microwave domains into THz imaging. Phase contrast [117] and darkfield imaging [118] have been demonstrated for unstained biological samples for observation of shape and density with better contrast. THz holograms have been recorded using highly coherent THz beams with much better resolution by appropriate amplitude and phase reconstruction, even with one-shot exposure [119–121]. Fourier imaging [122] and 3D vision [123] using the tomographic technique have improved tremendously in the last few years.

Overcoming the diffraction barrier has always been a challenge for researchers. Previously, near-field imaging methods using TDS (similar to scanning near-field optical microscopy, SNOM) have been successfully implemented in this domain by placing the object very close to the source with tens of micrometre spot size [124–126]. In recent years, the search for superresolution in the THz domain has taken centre stage by using new techniques. Techniques such as superresolution orthogonal deterministic imaging (SODI) are showing promising results [127]. Currently, we are witnessing a significant surge in this field with quality instrumentation being developed and used.

6 Conclusion

This review aimed at providing a brief introduction to a developing field of huge significance. The unique capabilities of THz waves have been discussed for many years, and now, they are becoming increasingly popular. Many commercial machines based on the principle of THz radiation are currently being incorporated into different walks of life such as security, pharmacy, NDT/NDE and biomedical imaging to name a few. Companies across the world are showing great interest in this field and the current (in 2022) market share of THz industries stands at a whopping $ 0.6 billion; projected to double within five years [128]. All these are signs of great progress for a field once referred to as a 'technology gap.'

References

1. D.M. Mittleman, Twenty years of terahertz imaging. Opt. Express **26**, 9417–9431 (2018)
2. Y.-S. Lee, *Principles of Terahertz Science* (Springer, 2009)
3. H. Elayan, O. Amin, B. Shihada, R.M. Shubair, M.S. Alouini, Terahertz band: the last piece of RF spectrum puzzle for communication systems. IEEE Open J. Commun. Soc. **1**, 1–32 (2020)
4. K. Ahi, N. Jessurun, M.-P. Hosseini, N. Asadizanjani, Survey of terahertz photonics and biophotonics. Opt. Eng. **59**(6), 061629 (2020)
5. G. Valušis, A. Lisauskas, H. Yuan, W. Knap, H.G. Roskos, Roadmap of terahertz imaging 2021. Sensors **21**, 4092 (2021)
6. Ch. Fattinger, D. Grischkowsky, Terahertz beams. Appl. Fhys. Lett. **54**, 490 (1989)
7. M. Tonouchi, Cutting-edge terahertz technology. Nat. Photonics **1**, 97–105 (2007)
8. P.H. Siegel, THz technology: an overview. Int. J. High Speed Electron. Syst. **13**(2), 351–394 (2003)
9. J.H. Son, Principle and applications of terahertz molecular imaging. Nanotechnology **24**, 214001 (2013)
10. B.B. Hu, M.C. Nuss, Imaging with terahertz waves. Opt. Lett. **20**, 1716 (1995)
11. M. Wan, J.J. Healy, J.T. Sheridan, Terahertz phase imaging and biomedical applications. Opt. Laser Technol. **122**, 105859–105871 (2020)
12. Q. Sun, Y. He, K. Liu, S. Fan, E. P. J. Parrott, E.Pickwell-MacPherson, Recent advances in terahertz technology for biomedical applications. Quant. Imaging. Med. Surg. **7**, 345–355 (2017)
13. M. Naftaly, N. Vieweg, A. Deninger, Industrial applications of terahertz sensing: state of play. Sensors **19**(19), 4203–4238 (2019)
14. F. Federici, B. Schulkin, F. Huang, D. Gary, R. Barat, F. Oliveira, D. Zimdars, THz imaging and sensing for security applications, explosives, weapons and drugs. Semicond. Sci. Technol. **20**, S266eS280 (2005)
15. A. D'Arco, M.D. Fabrizio, V. Dolci, A. Marcelli, M. Petrarca, G.D. Ventura, S. Lupi, Characterization of volatile organic compounds (VOCs) in their liquid-phase by terahertz time-domain spectroscopy. Biomed. Opt. Express **11**, 1–6 (2020)
16. Z. Chen et al., A survey on terahertz communications. China Commun. **16**(2), 1–35 (2019). https://doi.org/10.12676/j.cc.2019.02.001
17. A. Cosentino, Terahertz and cultural heritage science: examination of art and archaeology. Technologies **4**, 6–19 (2016)
18. P.R. Smith, D.H. Auston, M.C. Nuss, Subpicosecond photo conducting dipole antennas. IEEE J. Quantum Electron. **24**, 255 (1988)

19. Y.C. Shen, P.C. Upadhya, E.H. Linfield, H.E. Beere, A.G. Davies, Ultra-broadbandterahertz radiation from low-temperature-grown GaAs photoconductive emitters. Appl. Phys. Lett. **83**(15), 3117–3119 (2003)
20. A. Leitenstorfer, S. Hunsche, J. Shah, M.C. Nuss, W.H. Knox, Detectors and sources for ultra broadband electro-optic sampling: experiment and theory. Appl. Phys. Lett. **74**(11), 1516–1518 (1999)
21. B. Williams, Terahertz quantum cascade lasers. Nat. Photonics. **1**, 517 (2007)
22. L. Angrisani, G. Cavallo, A. Liccardo, G.P. Papari, A. Andreone, THz measurement systems, in *Book Chapter in New Trends and Developments in Metrology*, ed. by L. Cocco (IntechOpen, 2016)
23. D. Mittleman, Terahertz imaging, in *Book Chapter in Sensing with Terahertz Radiation*, ed. by D. Mittleman (Springer, Berlin, Heidelberg, 2003)
24. I.S. Gregory et al., Continuous-wave terahertz system with a 60 dB dynamic range. Appl. Phys. Lett. **86**, 204104 (2005)
25. K. Ahi, M. Anwar, Developing terahertz imaging equation and enhancement of the resolution of terahertz images using deconvolution. Proc. SPIE **9856**, 98560N (2016)
26. P. Lopato, T. Chady, Terahertz detection and identification of defects in layered polymer composites and composite coatings. Nondestruct. Test. Eval. **28**, 28–43 (2013)
27. N.S. Balbekin et al., Nondestructive monitoring of aircraft composites using terahertz radiation. Proc. SPIE **9448**, 94482D (2015)
28. T. Hagelschuer et al., High-spectral-resolution terahertz imaging with a quantum-cascade laser. Opt. Express **24**, 13839–13849 (2016)
29. E.V. Yakovlev et al., Non-destructive evaluation of polymer composite materials at theman-ufacturing stage using terahertz pulsed spectroscopy. IEEE Trans. Terahertz Sci. Technol. **5**, 810–816 (2015)
30. M. Kato et al., Non-destructive drug inspection in covering materials using a terahertz spectral imaging system with injection-seeded terahertz parametric generation and detection. Opt. Express **24**, 6425 (2016)
31. U.Schmidhammer, P. Jeunesse, Pulsed THz imaging for non-destructive testing of adhesive bonds, in *39th International Conference on Infrared, Millimeter, Terahertz Waves* (IEEE, 2014), pp. 1–2
32. X. Neiers, P. Jeunesse, U. Schmidhammer, Rapid control of machined glass fiber reinforced plastics by single shot terahertz time domain spectroscopy, in *40th Internal Conference on Infrared, Millimeter, Terahertz Waves*, vol. 13 (IEEE, 2015), pp. 1–2
33. K. Ahi, M. Anwar, Modeling of terahertz images based on x-ray images: a novel approach for verification of terahertz images and identification of objects with fine details beyond terahertz resolution. Proc. SPIE **9856**, 985610 (2016)
34. K. Ahi et al., Terahertz characterization of electronic components and comparison of terahertz imaging with x-ray imaging techniques. Proc. SPIE **9483**, 94830K (2015)
35. V.A. Trofimov et al., New algorithm for detection of dangerous objects hidden on a human body using passive THz camera. Proc. SPIE **9993**, 999305 (2016)
36. V.A. Trofimov et al., Concealed object detection using the passive THz image without its viewing. Proc. SPIE **9830**, 98300E (2016)
37. I.N. Dolganova et al., A hybrid continuous-wave terahertz imaging system. Rev. Sci. Instrum. **86**, 113704 (2015)
38. M. Kowalski, M. Kastek, Comparative studies of passive imaging in terahertz and mid-wavelength infrared ranges for object detection. IEEE Trans. Inf. Forensics Secur. **11**, 2028–2035 (2016)
39. M. Kowalski, Hidden object detection and recognition in passive terahertz and mid-wavelength infrared. J. Infrared Millim. Terahertz Waves **40**, 1074–1091 (2019)
40. P. Lopato, Estimation of layered materials dielectric parameters using pulsed terahertz technique. Int. J. Appl. Electromagn. Mech. **43**, 161–168 (2013)
41. P. Knobloch et al., Medical THz imaging: an investigation of histo-pathological samples. Phys. Med. Biol. **47**, 3875–3884 (2002)

42. O.A. Smolyanskaya et al., Glycerol dehydration of native and diabetic animal tissues studied by THz-TDS and NMR methods. Biomed. Opt. Express **9**, 1198 (2018)
43. O.A. Smolyanskaya et al., Terahertz biophotonics as a tool for studies of dielectric and spectral properties of biological tissues and liquids. Prog. Quantum Electron. **62**, 1–77 (2018)
44. A.A. Gavdush et al., Terahertz spectroscopy of gelatin-embedded human brain gliomas of different grades: a road toward intraoperative THz diagnosis. J. Biomed. Opt. **24**, 027001 (2019)
45. E.A. Strepitov, et al., Analysis of spectral characteristics of normal fibroblasts and fibroblasts cultured with cancer cells in terahertz frequency range, in *Progress In Electromagnetics Research Symposium [proceedings]* (2014), pp. 1707–1710
46. M.V. Duka et al., Numerical and experimental studies of mechanisms underlying the effect of pulsed broadband terahertz radiation on nerve cells. Quantum Electron. **44**, 707–712 (2014)
47. K.I. Zaytsev et al., Highly accurate in vivo terahertz spectroscopy of healthy skin: variation of refractive index and absorption coefficient along the human body. IEEE Trans. Terahertz Sci. Technol. **5**, 817–827 (2015)
48. M.H. Arbab et al., Terahertz spectroscopy for the assessment of burn injuries in vivo. J. Biomed. Opt. **18**, 077004 (2013)
49. J.A. Zeitler, D.A. Newnham, P.F. Taday, T.L. Threlfall, R.W. Lancaster, R.W. Berg, C.J. Strachan, M. Pepper, K.C. Gordon, T. Rades, Characterization of temperature-induced phase transitions in five polymorphic forms of sulfathiazole by terahertz pulsed spectroscopy and differential scanning calorimetry, J. Pharm. Sci. **95**, 2486e2498 (2006)
50. C.J. Strachan, T. Rades, D.A. Newnham, K.C. Gordon, M. Pepper, P.F. Taday, Using terahertz pulsed spectroscopy to study crystallinity of pharmaceutical materials. Chem. Phys. Lett. **390**, 20e24 (2004)
51. M. Yamaguchi, F. Miyamaru, K. Yamamoto, M. Tani, M. Hangyo, Terahertz absorption spectra of L-, D-, and DL-alanine and their application to determination of enantiometric composition. Appl. Phys. Lett. **86**, 1e3 (2005)
52. J.A. Zeitler, K. Kogermann, J. Rantanen, T. Rades, P.F. Taday, M. Pepper, J. Aaltonen, C.J. Strachan, Drug hydrate systems and dehydration processes studied by terahertz pulsed spectroscopy. Int. J. Pharm. **334**, 78e84 (2007)
53. H.B. Liu, X.C. Zhang, Dehydration kinetics of D-glucose monohydrate studied using THz time-domain spectroscopy. Chem. Phys. Lett. **429**, 229e233 (2006)
54. K.L. Nguyen, T. Friscic, G.M. Day, L.F. Gladden, W. Jones, Terahertz timedomain spectroscopy and the quantitative monitoring of mechanochemical cocrystal formation. Nat. Mater. **6**, 206e209 (2007)
55. E.P.J. Parrott, J.A. Zeitler, T. Friscic, M. Pepper, W. Jones, G.M. Day, L.F. Gladden, Testing the sensitivity of terahertz spectroscopy to changes in molecular and supramolecular structure: a study of structurally similar cocrystals. Cryst. Growth Des. **9**, 1452e1460 (2009)
56. Z. Popovic, E.N. Grossman, THz metrology and instrumentation. IEEE Trans. Terahertz Sci. Technol. **1**(1), 133–144 (2011)
57. G. Kozlov, A. Volkov, *Coherent Source Submillimeter Wave Spectroscopy in Millimeter and Submillimeter Wave Spectroscopy of Solid* (Springer, Berlin, 1998), pp.51–109
58. L. Qian, S.D. Benjamin, P.W.E. Smith, B.J. Robinson, D.A. Thompson, Picosecond carrier lifetime and large optical nonlinearities in InGaAsP grown by He-plasma-assisted molecular beam epitaxy. Opt. Lett. **22**, 108 (1997)
59. T. Kondo, M. Sakamoto, M. Tonouchi, M. Hangyo, Terahertz radiation from (111) InAs surface using 1.55 11m femtosecond laser pulses. Jpn. J. Appl. Phys. Pt. 2 (Lptt.) **38**, 1035 (1999)
60. N. Sekine, K. Hirakawa, F. Sogawa, Y. Arakawa, N. Usami, Y. Shiraki, T. Katoda, Ultrashort lifetime photocarriers in Ge thin films. Appl. Phys. Lett. **68**, 3419 (1996)
61. N. Katzenellenbogen, D. Grischkowsky, Efficient generation of 380 fs pulses of THz radiation by ultrafast laser pulse excitation of a biased metalsemiconductor interface. Appl. Phys. Lett. **58**, 222 (1991)

62. S.E. Ralph, D. Grischkowsky, THz spectroscopy and source characterization by optoelectronic interferometry. Appl. Phys. Lett. **60**, 1070 (1992)
63. F. Sizov, Terahertz radiation detectors: the state-of-theart. Semicond. Sci. Technol. **33**, 123001–123027 (2018)
64. T.A. Liu, M. Tani, M. Nakajima, M. Hangyo, C.L. Pan, Ultra-broadband terahertz field detection by photoconductive antennas based on multi-energy arsenic ion implanted GaAs and semi-insulating GaAs. Appl. Phys. Lett. **83**(7), 1322 (2003)
65. S. Kono, M. Tani, K. Sakai, Ultra-broadband photoconductive detection: comparison with free-space electro-optical sampling. Appl. Phys. Lett. **79**(7), 898–900 (2001)
66. R.B. Kohlhaas et al., Photoconductive terahertz detectors with 105 dB peak dynamic range made of rhodium doped InGaAs. Appl. Phys. Lett. **114**, 221103–221109 (2019)
67. Y. Salamin et al., Compact and ultra-efficient broadband plasmonic terahertz field detector. Nat. Commun. **10**, 5550 (2019)
68. P. Jepsen, R.H. Jacobsen, S.R. Keiding, Generation and detection of terahertz pulses from biased semiconductor antennas. J. Opt. Soc. Am. B **13**, 2424 (1996)
69. D.M. Mittleman, R.H. Jacobsen, M.C. Nuss, T-ray imaging. IEEE J. Sel. Top. Quantrum Electron. **2**, 679 (1996)
70. J.V. Rudd, D.M. Mittleman, The influence of substrate lens design in terahertz time-domain spectroscopy. J. Opt. Soc. Am. B **19**, 319 (2002)
71. D.F. Filipovic, S.S. Gearhart, G.M. Rebeiz, Double-slot antennas on extended hemispherical and elliptical silicon dielectric lenses. IEEE Trans. Microwave Theory Tech. **41**, 1738 (1993)
72. W.B. Dou, G. Zeng, Z.L. Sun, Pattern prediction of extended hemispherical lens/objective lens antenna system at millimeter wavelengths. IEE Proc. Microwave Antennas Propag. **145**, 295 (1998)
73. J. R. Bray, L. Roy, Performance trade-offs of substrate lens antennas, in *Proceedings of the Symposium on Antenna Technology and Applied Electromagnetics (ANTEM), Ottawa, Canada* (1998), pp. 321–324
74. A. Siemion, Terahertz diffractive optics—smart control over radiation. J. Infrared Millim. Terahertz Waves **40**, 477–499 (2019)
75. A. Siemion, The magic of optics—an overview of recent advanced terahertz diffractive optical elements. Sensors **21**, 100 (2021)
76. D. Headland, Y. Monnai, D. Abbott, C. Fumeaux, W. Withayachumnankul, Tutorial: terahertz beamforming, from concepts to realizations. APL Photonics **3**, 051101 (2018)
77. R. Ivaškevičiūtė-Povilauskienė, L. Minkevičius, D. Jokubauskis, A. Urbanowicz, S. Indrišiūnas, G. Valušis, Flexible materials for terahertz optics: advantages of graphite-based structures. Opt. Mater. Express **9**, 4438–4446 (2019)
78. L.D. VanPutten, J. Gorecki, E. Numkam Fokoua, V. Apostolopoulos, F. Poletti, 3D-printed polymer antiresonant waveguides for short-reach terahertz applications. Appl. Opt. **57**, 3953–3958 (2018)
79. J. Bomba, J. Suszek, M. Makowski, A. Sobczyk, M. Sypek, 3-D printed anti-reflection structures for the terahertz region. J. InfraredMillim. Terahertz Waves **39**, 24–35 (2018)
80. E.D. Walsby, S. Wang, J. Xu, T. Yuan, R. Blaikie, S.M. Durbin, X.C. Zhang, D.R. Cumming, Multilevel silicon diffractive optics for terahertz waves. J. Vac. Sci. Technol. Microelectron. Nanometer Struct. **20**, 2780–2783 (2002)
81. L. Minkevičius, S. Indrišiūnas, R. Šniaukas, B. Voisiat, V. Janonis, V. Tamošiūnas, I. Kašalynas, G. Račiukaitis, G. Valušis, Terahertz multilevel phase Fresnel lenses fabricated by laser patterning of silicon. Opt. Lett. **42**, 1875–1878 (2017)
82. D.K. Gramotnev, S.I. Bozhevolnyi, Plasmonics beyond the diffraction limit. Nat. Photonics **4**, 83–91 (2010)
83. J.A. Schuller, E.S. Barnard, W. Cai, Y.C. Jun, J.S. White, M.L. Brongersma, Plasmonics for extreme light concentration and manipulation. Nat. Mater. **9**, 193–204 (2010)
84. R. Dharmavarapu, K.I. Izumi, I. Katayama, S.H. Ng, J. Vongsvivut, M.J. Tobin, A. Kuchmizhak, Y. Nishijima, S. Bhattacharya, S. Juodkazis, Dielectric cross-shaped resonator based metasurface for vortex beam generation in Mid-IR and THz wavelengths. De Gruyter **8**, 1263–1270 (2019)

85. K. Fan, J. Zhang, X. Liu, G.F. Zhang, R.D. Averitt, W.J. Padilla, Phototunable dielectric huygens' metasurfaces. Adv. Mater. **30**, 1800278 (2018)
86. X. You, R.T. Ako, W.S. Lee, M. Bhaskaran, S. Sriram, C. Fumeaux, W. Withayachumnankul, Broadband terahertz transmissive quarter-wave metasurface. APL Photonics **5**, 096108 (2020)
87. X. Zang, C. Mao, X. Guo, G. You, H. Yang, L. Chen, Y. Zhu, S. Zhuang, Polarization-controlled terahertz super-focusing. Appl. Phys. Lett. **113**, 071102 (2018)
88. J. Dong, A. Locquet, M. Melis, D.S. Citrin, Global mapping of stratigraphy of an old-master painting using sparsity-based terahertz reflectometry. Sci. Rep. **7**, 15098 (2017)
89. H. Guerboukha, K. Nallappan, M. Skorobogatiy, Exploiting k-space/frequency duality toward real-time terahertz imaging. Optica **5**, 109–116 (2018)
90. M. Qiao, Z. Meng, J. Ma, X. Yuan, Deep learning for video compressive sensing. APL Photonics **5**, 030801 (2020)
91. T.S. Hartwick, D.T. Hodges, D.H. Barker, F.B. Foote, Far infrared imagery. Appl. Opt. **1976**, 15 (1919)
92. S.S. Dhillon, M.S. Vitiello, E.H. Linfield, A.G. Davies, M.C. Hoffmann, J. Booske, C. Paoloni, M. Gensch, P. Weightman, G.P. Williams et al., The 2017 terahertz science and technology roadmap. J. Phys. Appl. Phys. **50**, 043001 (2017)
93. H. Zhang, S. Sfarra, A. Osman, K. Szielasko, C. Stumm, M. Genest, X.P. Maldague, An infrared-induced terahertz imaging modality for foreign object detection in a lightweight honeycomb composite structure. IEEE Trans. Ind. Inform. **14**, 5629–5636 (2018)
94. A.O. Chulkov, S. Sfarra, H. Zhang, A. Osman, K. Szielasko, C. Stumm, F. Sarasini, J. Fiorelli, X.P. Maldague, V.P. Vavilov, Evaluating thermal properties of sugarcane bagasse-based composites by using active infrared thermography and terahertz imaging. Infrared Phys. Technol. **97**, 432–439 (2019)
95. A. Bendada, S. Sfarra, C. Ibarra-Castanedo, M. Akhloufi, J.P. Caumes, C. Pradere, J.C. Batsale, X. Maldague, Subsurface imaging for panel paintings inspection: a comparative study of the ultraviolet, the visible, the Infrared and the terahertz spectra. Opto-Electron. Rev. **23**, 88–99 (2015)
96. H. Zhang, S. Sfarra, K. Saluja, J. Peeters, J. Fleuret, Y. Duan, H. Fernandes, N. Avdelidis, C. Ibarra-Castanedo, X. Maldague, Non-destructive investigation of paintings on canvas by continuous wave terahertz imaging and flash thermography. J. Nondestruct. Eval. **36**, 34 (2017)
97. J.T. Kindt, C.A. Schmuttenmaer, Far-infrared dielectric properties of polar liquids probed by femtosecond terahertz pulse spectroscopy. J. Phys. Chem. **100**, 10373 (1996)
98. C. Ronne, L. Thrane, P.-O. Astrand, A. Wallqvist, K.V. Mikkelsen, S.R. Keiding, Investigation of the temperature dependence of dielectric relaxation in liquid water by THz reflection spectroscopy and molecular dynamics simulation. J. Chem. Phys. **107**, 5319 (1997)
99. L. Thrane, R.H. Jacobsen, P. Uhd Jepsen, S.R. Keiding, THz reflection spectroscopy of liquid water. Chem. Phys. Lett. **240**, 330 (1995)
100. M. Sadeghi, M.M. Tajdini, E. Wig, C.M. Rappaport, Single-frequency fast dielectric characterization of concealed body-worn explosive threats. IEEE Trans. Antennas Propag. **68**, 7541–7548 (2020)
101. A. Dobroiu, C. Otani, K. Kawase, Meas. Sci. Technol. **17**, R161 (2006)
102. A.W.M. Lee, Q. Hu, Opt. Lett. **30**, 2563 (2005)
103. I.S. Gregory, W.R. Tribe, C. Baker, B.E. Cole, M.J. Evans, L. Spencer, M. Pepper, M. Missous, Appl. Phys. Lett. **86**, 204104 (2005)
104. D. Zimdars, G. Stuk, S Williamson of Picometrix LLC. Terahertz imaging in reflection and transmission mode for luggage and personnel inspection. US Provisional Application No. 60/574.643 filed May 26, 2004
105. A. Ingar Romero, A.K. Mukherjee, A. Fernandez Olvera, et al., Visualizing nanometric structures with sub-millimeter waves. Nat. Commun.**12**, 7091 (2021)
106. A.W.M. Lee, Q. Qin, S. Kumar, B.S. Williams, Q. Hu, J.L. Reno, Appl. Phys. Lett. **89**, 141125 (2006)
107. I.S. Gregory, W.R. Tribe, C. Baker, B.E. Cole, M.J. Evans, L. Spencer, M. Pepper, M. Missous, Appl. Phys. Lett. **86**, 204104 (2005)

108. N. Karpowicz, et al., Comparison between pulsed terahertz time-domain imaging and continuous wave terahertz imaging. Semicond. Sci. Technol. **20**, S293 (2005)
109. N. Dolganova et al., Nanoparticle-enabled experimentally trained wavelet-domain denoising method for optical coherence tomography. J. Biomed. Opt. **23**, 091406 (2018)
110. B. Ferguson et al., T-ray computed tomography. Opt. Lett. **27**, 1312 (2002)
111. N. Rothbart et al., Fast 2-D and 3-D terahertz imaging with a quantum-cascade laser and a scanning mirror. IEEE Trans. Terahertz Sci. Technol. **3**, 617–624 (2013)
112. M.W. Ayech, D. Ziou, Terahertz image segmentation using k-means clustering based on weighted feature learning and random pixel sampling. Neurocomputing **175**, 243–264 (2016)
113. H. Zhang et al., Numerical and experimental analyses for natural and non-natural impacted composites via thermographic inspection, ultrasonic C-scan and terahertz imaging. Proc. SPIE **10214**, 102140I (2017)
114. P. Lopato, Pulsed excitation terahertz tomography—multiparametric approach. Open Phys. **16**, 111–116 (2018)
115. A. Redo-Sanchez, B. Heshmat, A. Aghasi et al., Terahertz time-gated spectral imaging for content extraction through layered structures. Nat Commun **7**, 12665 (2016)
116. D.M. Mittleman, S. Hunsche, L. Boivin, M.C. Nuss, T-ray tomography. Opt. Lett. **22**, 904–906 (1997)
117. Z. Liu, L. Tian, S. Liu, L. Waller, Real-time brightfield, darkfield, and phase contrast imaging in a light-emitting diode array microscope. J. Biomed. Opt. **19**, 106002 (2014)
118. T. Löffler, T. Bauer, K.J. Siebert, H.G. Roskos, A. Fitzgerald, S. Czasch, Terahertz dark-field imaging of biomedical tissue. Opt. Express **9**, 616–621 (2001)
119. L. Valzania, P. Zolliker, E. Hack, Coherent reconstruction of a textile and a hidden object with terahertz radiation. Optica **6**, 518–523 (2019)
120. N.V. Petrov, M.S. Kulya, A.N. Tsypkin, V.G. Bespalov, A. Gorodetsky, Application of terahertz pulse time-domain holography for phase imaging. IEEE Trans. Terahertz Sci. Technol. **6**, 464–472 (2016).
121. M. Locatelli, M. Ravaro, S. Bartalini, L. Consolino, M.S. Vitiello, R. Cicchi, F. Pavone, P. De Natale, Real-time terahertz digital holography with a quantum cascade laser. Sci. Rep. **5**, 13566 (2015)
122. H. Yuan, D. Voß, A. Lisauskas, D. Mundy, H.G. Roskos, 3D Fourier imaging based on 2D heterodyne detection at THz frequencies. APL Photonics **4**, 106108 (2019)
123. J.B. Perraud, J.P. Guillet, O. Redon, M. Hamdi, F. Simoens, P. Mounaix, Shape-from-focus for real-time terahertz 3D imaging. Opt. Lett. **44**, 483–486 (2019)
124. F. Keilmann, FIR microscopy. Infrared Phys. Technol. **36**, 217–224 (1995)
125. T. Pfeifer, H.M. Heiliger, T. Löffler, C. Ohlhoff, C. Meyer, G. Lüpke, H.G. Roskos, H. Kurz, Optoelectronic on-chip characterization of ultrafast electric devices: Measurement techniques and applications. IEEE J. Sel. Top. Quantum Electron. **2**, 586–604 (1996)
126. J.R. Knab, A.J.L. Adam, E. Shaner, H.J.A.J. Starmans, P.C.M. Planken, Terahertz near-field spectroscopy of filled subwavelength sized apertures in thin metal films. Opt. Express **21**, 1101–1112 (2013)
127. H. Guerboukha, Y. Cao, K. Nallappan, M. Skorobogatiy, Super-Resolution Orthogonal Deterministic Imaging Technique for Terahertz Subwavelength Microscopy. ACS Photonics **7**, 1866–1875 (2020)
128. Researchandmarkets.com, Global Terahertz Technology Market (2022–2027), a Report

Design of Super-Achromatic Phase Controlling Assemblies for THz Spectro-Polarimetric Imaging System Using Metaheuristic Optimization Technique

Nilanjan Mukhopadhyay and Arijit Saha

1 Introduction

A quarter wave phase retarder (QWR) is responsible for generating circularly polarized light. Their performance usually depends on the wavelength that can be nullified following achromatic designing. An achromatic phase retarder (or waveplate) is a crucial component of any imaging system that takes care of the polarization states of incoming light. Very often these achromatic phase retarders are associated with different birefringent optical filters in various scientific and engineering applications. Designing achromatic phase retarders has been a domain of interest for optical system designers over the years. Birefringent materials play an important role when such achromatic designing is required. Pancharatnam showed the way with his achromatic combination of the same birefringent materials over a visible wavelength range [1]. Later on, researchers proposed systems with crystalline quartz as well as sapphire as birefringent materials [2, 3]. Since the rotation angle strongly depends on wavelength, a combination of different birefringent materials without fast axis rotation has been proposed to overcome the wavelength dependency [4]. Nowadays, near-infrared (NIR) to short-wave infrared (SWIR) range of electromagnetic spectrum comes up with different applications, e.g., remote sensing, spectroscopy and medical imaging [5–11]. A super-achromatic system having quartz and MgF_2 waveplates has been proposed operating over visible to NIR range [12]. A multilayer achromatic waveplate operating over 1000–1800 nm has also been proposed that can be used in a Infrared Imaging Magnetograph (IRIM) system [13]. Several

N. Mukhopadhyay (✉)
Department of Electronics and Communication Engineering, Global Institute of Management and Technology, Krishnagar, Nadia 741102, India
e-mail: nilu.opt@gmail.com

A. Saha
Department of Electronics and Communication Engineering, B. P. Poddar Institute of Management and Technology, 137, VIP Road, Kolkata 700 052, India

A. Saha et al. (eds.), *Optical to Terahertz Engineering*, Progress in Optical Science and Photonics 23, https://doi.org/10.1007/978-981-99-0228-6_2

other multi-crystal achromatic and super-achromatic systems have been designed and characterized over NIR to SWIR range [14–16]. There are several other techniques used for designing achromatic phase retarders. Achromatic phase retarder using sub-wavelength dielectric diffraction grating (SWGs) has been studied by Nicolas Passily et al. and H. Kikuta et al. in both the visible and near-infrared regions [17, 18]. These types of designs have been reported with a retardance error smaller than 3°. Achromatic phase retarders in mid-wave infrared (MWIR) and low-wave infrared (LWIR) have also been proposed with SWGs [19, 20]. A metal-multilayer dielectric grating has been reported as having a retardance deviation of 90° ± 4° operating over a near-infrared range [21]. The main drawbacks of these kinds of systems are associated with high designing complexity as the overall performance is sensitive to the period of the grating structure and the orientation angle. Moreover, liquid crystals (LC), the main driving component of several display applications, are now being used in designing infrared phase retarders, which operate with very small voltage [22–24].

In recent years, the world has seen a wide range of applications with THz techniques such as terahertz time-domain spectroscopy (THz-TDS) [25], biological, medical and pharmaceutical sciences [26–31], environmental hazards and security [32–34], explosives inspection [35], semiconductor industry [36, 37], Information and Communication Technology (ICT) sector [38, 39] and many more. Over the past few years, researchers have reported different achromatic retarders in the THz range used in imaging and detection. An achromatic THz phase retarder operating over 0.25 to 1.75 THz has been designed with 6 birefringent quartz plates [40]. A simulated annealing algorithm has been used to optimize thicknesses (31.4 mm) and relative orientations to achieve a phase deviation of 90° ± 3°%. Another system has been proposed with 9 pieces of crystalline quartz plates having different thicknesses and different rotation angles over 1.3–1.8 THz [41]. They have managed to achieve a phase delay of 90° ± 3° with a total thickness of around 52 mm. An achromatic quarter waveplate based on silicon grating has been proposed with ±3% phase variation over the 0.47–0.8 THz range [42]. Hsieh et al. have designed a tunable achromatic quarter-wave phase retarder based on liquid crystal (LC) with a phase deviation of 90° ± 9° from 0.20 to 0.50 THz [43].

Recently, Multiangle Spectro-Polarimetric Imager (MSPI) is being used for the Aerosol-Cloud-Ecosystem (ACE) mission by NASA both for ground-based and airborne cameras [44]. A three plate (Sapphire; MgF_2; Quartz) achromatic quarter-wave phase retarder found its application in the system with retardance variation of 90° ± 10°. A multi-crystal achromatic quarter-wave phase retarder with improved achromatic behaviour for the MSPI system has also been proposed [45]. In this present communication, a three-waveplate super-achromatic quarter-wave retarder (SQWR) combination has been designed and characterized over a 1.7–2.3 μm wavelength spectrum. The 'super-achromatic quarter-wave phase retarder' consists of calcite, crystalline quartz and ADP in a sequential combination of negative–positive-negative waveplates. Performance of the retarder with respect to the change of temperature and waveplate thickness has been elaborated, and it exhibits good achromatic behaviour over the intended wavelength spectrum.

In the very next section, i.e., Sect. 2, theory and designing methodology have been illustrated which is followed by the internal structure of the MSPI system in Sect. 2.1. Section 3 deals with simulation results and discussion, and the overall transmittance of the system is discussed in Sect. 4. Finally, a conclusion has been drawn in Sect. 5.

2 Theoretical Analysis

Phase difference acquired by the light beam while passing through the birefringent plates depends on optical path difference according to the following equation [14]:

$$\varphi_i = \frac{2\pi f}{c} \sum_{j=1}^{M} \Delta n_{i,j} d_j \tag{1}$$

where φ_i is overall phase retardation of an M-plate optical retarder system operating with design frequency, i.e., f_i. Birefringence of the j-th plate (with a thickness of the materials d_j) is $\Delta n_{i,j,}$ and f is the overall frequency spectrum of interest. Optical path difference is calculated as the product of birefringence and thickness of the waveplate, and for quarter-wave phase retarder (i.e., phase difference of π/2), it is equal to ¼th of wavelength. This relation can be presented in a generalized matrix form considering an optical system of M waveplates with thickness given as d_1, d_2, ..., d_M respectively having N number of design frequencies, i.e., f_1, f_2, ..., f_N:

$$\begin{vmatrix} \Delta n_{11} & \Delta n_{12} & .. & \Delta n_{1M} \\ \Delta n_{21} & \Delta n_{22} & .. & \Delta n_{2M} \\ .. & .. & .. & .. \\ \Delta n_{N1} & \Delta n_{N2} & .. & \Delta n_{NM} \end{vmatrix} \begin{vmatrix} d_1 \\ d_2 \\ .. \\ d_M \end{vmatrix} = \pm\frac{c}{4} \begin{vmatrix} f_1^{-1} \\ f_2^{-1} \\ .. \\ f_N^{-1} \end{vmatrix} \tag{2}$$

where birefringence of M-th plate is Δn_{NM} at design frequency f_N. From Eq. 2, thicknesses of the plates can be computed using Cramer's rule [15].

2.1 Designing Steps

Designing SQWR with different birefringent materials is based on thickness calculation at particular design wavelengths. This designing methodology has the advantage of no fast axis rotation of waveplates as compared to that of the same birefringent material-based systems. Therefore, the dependency of fast axis rotation on wavelength can be overcome. The proposed system consists of three birefringent materials, namely calcite, crystalline quartz and ADP.

Thicknesses of the three waveplates can be computed by solving Eq. (2) considering birefringence values for the materials at three design frequencies, i.e., $f_1 =$

133.2 THz, $f_2 = 150$ THz and $f_3 = 176.5$ THz respectively selected over 133–176 THz. Variations of birefringence of all three materials are illustrated in Fig. 1. Birefringence values of the three crystals are taken from [46]. Table 1 illustrates the obtained thickness values.

In the case of SQWR, the wavelength dependency of overall phase retardation has been reduced. Constant retardation value over the intended frequency region can be achieved by selecting a suitable waveplate thickness that can be considered as a nonlinear optimization problem. One of the recently proposed metaheuristic optimization algorithms known as the Flower pollination algorithm (FPA) [47], which is inspired by the pollination of flowers, has been taken into account for solving the problem. The square of the error between target retardation and obtained retardation is defined as objective function k for the FPA, written as follows:

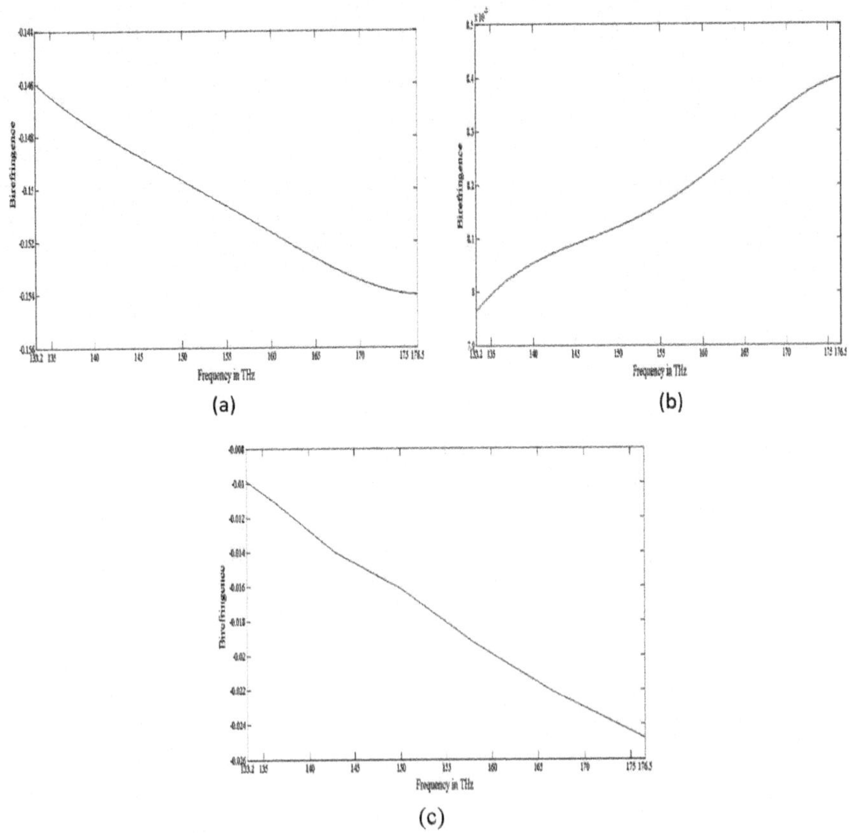

Fig. 1 Variation of birefringence throughout the intended frequency spectrum for **a** calcite, **b** crystalline quartz and **c** ADP waveplates

Table 1 Thickness values of the three waveplates

Name of the plates	Thickness value (μm)
Calcite (d_X)	2.9
Crystalline quartz (d_Y)	140.1
ADP (d_Z)	12.1

Table 2 Optimized thickness values of the three waveplates

Name of the plates	Thickness value (μm)
Calcite (d_X)	1.0
Crystalline quartz (d_Y)	103.7
ADP (d_Z)	11.7

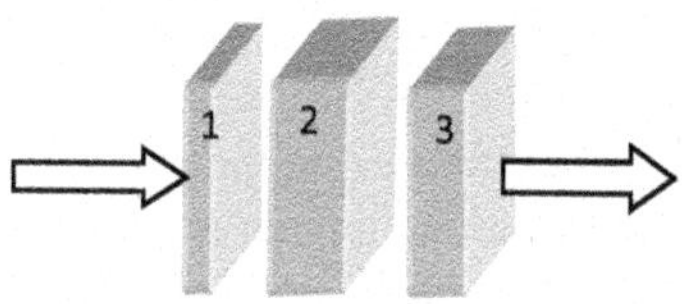

Fig. 2 Schematic diagram of the proposed three-plate super-achromatic quarter-wave phase retarder consisting of calcite (1), crystalline quartz (2) and ADP (3) materials in sequential order of thicknesses

$$k = \sum_{i=1}^{N} (\varphi_{Exp} - \varphi_{Cal,i})^2 \tag{3}$$

where target retardation is φ_{Exp}, which is π/2 in this case (quarter-wave retarder) and calculated retardation at design frequency f_i is $\varphi_{Cal,i}$. The optimum set of thickness values that minimizes the error function k is represented in Table 2.

Construction of the proposed system is shown in Fig. 2, where fast axes of each waveplate are mutually parallel, and they have been placed as negative (Calcite), positive (Quartz) and negative (ADP) birefringent material with respect to wavelength having variable thickness.

Over the intended frequency range, all three materials have high transmittance, low hygroscopicity, high melting point and low brittleness properties that come in support of their good candidature [48–50]. Waveplates of three birefringent materials are commercially well available.

3 Simulation Results and Discussion

In this section, the behaviour of the proposed three-plate system has been illustrated. Simulation results for the proposed system over the 133–176 THz range have been studied, and system performance is also compared with commercially available systems operating in the same frequency range. Among different optimization techniques available with different accuracies, FPA has been selected for this purpose because of its high efficiency and fast converging speed [47]. The following specifications have been considered in the algorithm, e.g., the number of iterations has been set to 200 along with population size and the probability switch values as 50 and 0.8, respectively. All the simulations are performed over MATLAB 2009a in a laptop with 2 GB RAM, Dual Core processor and Windows 7 operating system, and the time required for iteration is 5 s.

The effect of using three different birefringent materials has been shown in Fig. 3 where birefringence dispersion of the previous material is minimized by the next material and so on. Considering the thickness of all the waveplates as computed in Table 2, the combined retardance over the 133–176 THz range is obtained and illustrated in Fig. 3. Deviation of retardance from the target value (i.e., 90°) is within ±0.4°. Retardation deviation is computed as ±0.4° that indicates improvement in the achromatic behaviour of the system, and the results are in support of the system being referred to as 'super-achromatic'. Moreover, the percentage deviation of retardance is within +0.5% and −0.38% over the domain of interest.

There are some key design constraints that can have a huge effect on the super-achromatic behaviour of the proposed SQWR. These parameters have to be taken care of by an optical system designer. One of those such important parameters is retardation variation with a thickness of plates since phase difference is related to path difference. Figure 4 represents the retardation deviation with variation of waveplate thickness. Therefore, thickness tolerance for all the waveplates is crucial in this regard. Change in total retardation is within ±1% if the thickness of the Calcite plate

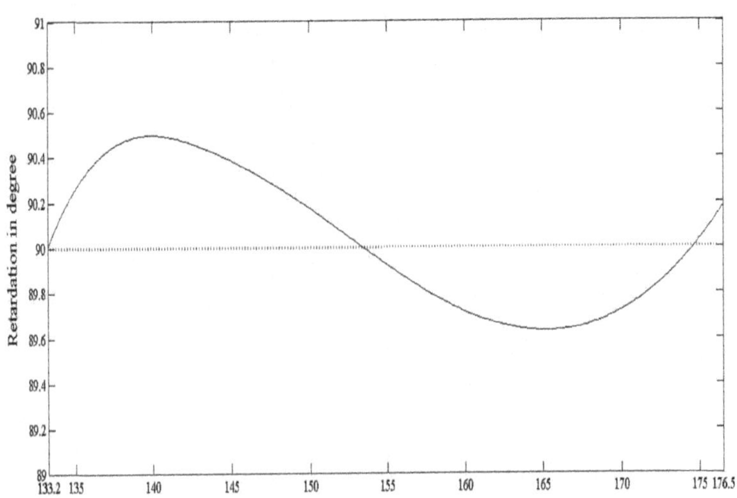

Fig. 3 Retardation variation of combined system over intended frequency range with optimized plate thickness

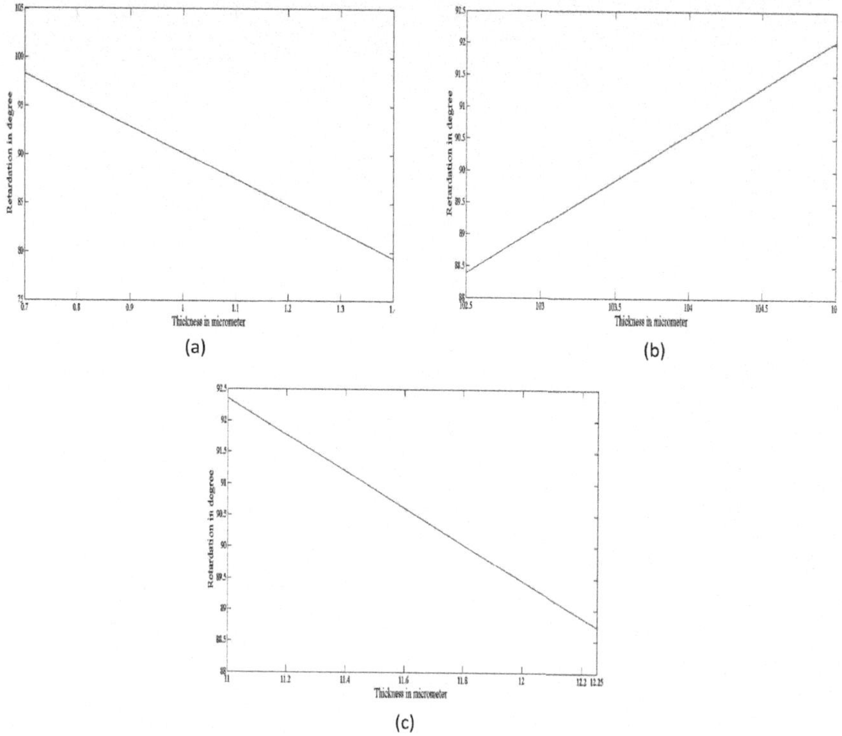

Fig. 4 Retardation variation with change in individual thickness of **a** calcite, **b** crystalline quartz and **c** ADP waveplates

varies from −15% to +10% from the computed thickness value. This thickness change for crystalline quartz is from −0.48% to +0.38% and that of ADP is within −0.85% to +2.56%. Although the thickness tolerance range is high for the Calcite plate, more precisive polishing during manufacturing is needed as the thickness of Calcite is very small compared to other plates. This study reflects that the proposed SQWR has significant phase stability with respect to thickness variation.

The very next constraint is the effect of temperature that eventually is another important aspect of the SQWR design. With the change in temperature, birefringence of the materials varies which in turn shifts overall phase retardation. The relation between refractive index and temperature for specific wavelength is given by [51]

$$n = n_o + \frac{dn}{dT}(T - T_o) \tag{4}$$

where n_0 is the refractive index at room temperature, T_0 is room temperature in Kelvin, T is the new temperature and dn/dT is the temperature coefficient of the refractive index. Values of dn/dT that depend on frequency are taken from optical handbooks [50]. Considering the temperature coefficient of refractive index values at

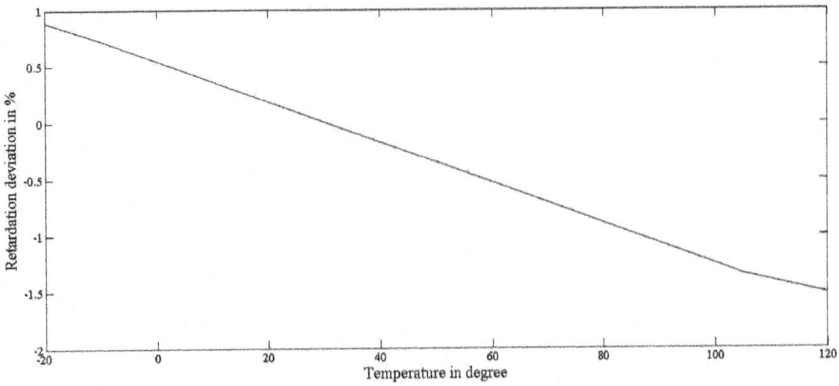

Fig. 5 Percentage deviation of retardation with temperature change

Table 3 Overall transmittance of the system

Frequency (THz)	Overall transmittance (%)
1.7	64.5
1.9	64.3
2.0	64.52

1.7 μm for all three materials, the retardation variation against temperature change has been illustrated in Fig. 5 following Eq. 4.

It can be observed that the operating temperature range where the retardation varies by 1.5% is from −20 to 120 °C. It indicates a wide range of operating temperatures for achromatic behaviour.

The overall transmittance of the system considering individual birefringent material's effect over the intended frequency spectrum has been illustrated in Table 3. Transmittance has been computed at three design frequencies and it is around 65%.

4 Applications

QWR is an important component in Spectro-polarimetric imaging systems that have a wide range of applications including aerosol and atmospheric studies and the biomedical domain, etc. This proposed SQWR can be used in Air Multiangle Spectro-Polarimetric Imager (AirMSPI) for airborne cameras applicable for the Aerosol-Cloud-Ecosystem (ACE) mission by NASA over NIR to SWIR wavelength range which belongs to THz frequency as shown in Fig. 6.

There are three curved mirrors to form off-axis astigmat, and the two photoelastic modulators (PEMs) are situated between two achromatic quarter waveplates. On the focal plane, the assembly of filters and polarizers are integrated [44]. Polarization errors due to retardance deviation produced by quarter-wave retarders degrades

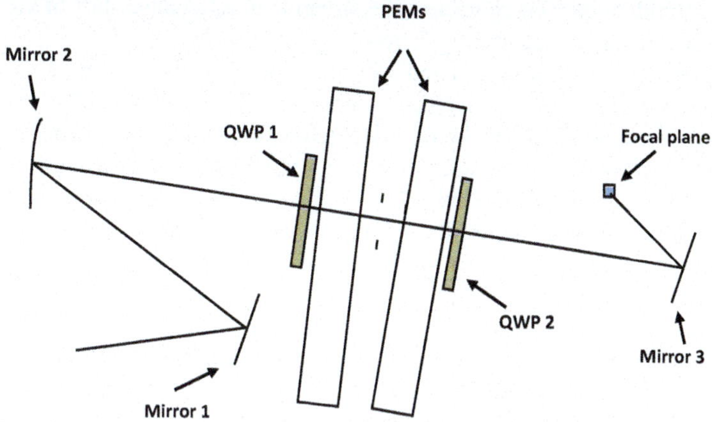

Fig. 6 Schematic diagram of AirMSPI optical design with two QWRs [25]

the signal-to-noise ratio (SNR). The proposed SQWR with retardation deviation of ±0.4° can be a suitable candidate as it will improve SNR. Moreover, the wide operating temperature range and excellent mechanical properties of the used birefringent materials make it applicable in different environmental conditions. Two versions of MSPI are currently in operation. GroundMSPI is a portable ground-based instrument operating at the University of Arizona's College of Optical Sciences. AirMSPI, a similar instrument with modifications suitable for aircraft flights, operates at JPL. AirMSPI has made two successful ER-2 flights, and more are planned. Both GroundMSPI and AirMSPI operate from the UV to the near IR, with polarimetric channels at 470, 660 and 865 nm and additional intensity-only channels at 355, 380, 445, 555 and 935 nm. Both instruments are fully operational, providing polarimetric images typically at 1536 × 5097 pixel resolution. The applications of MSPI extend beyond the needs of climate science. A partial list of topics for which MSPI spectro-polarimetric imagery can be useful includes Aerosol and atmospheric studies, Test bed for surface scattering studies, Test bed for object identification studies, Classification of objects and estimation of cardinal orientation, Dependence of polarization/scattering on wavelength and angle, Imaging through haze, Imaging through foliage, Imaging through water, Identification of sub-resolution objects due to anomalous polarization signatures, and Polarimetric scattering models of natural versus manmade scenes.

5 Conclusion

In this communication, a quarter-wave retarder that exhibits super-achromatic behaviour has been designed and characterized over the 133–176 THz range. Present

SQWR exhibits improved characteristics in terms of retardance shift of 90° ± 0.4° as compared to earlier reported system having 90° ± 10° shift.

The proposed SQWR is reported with three birefringent material combination that exhibits less retardance variation as compared to any commercially available systems. Moreover, the SQWR has good retardance stability with respect to thickness variations, temperature changes and also provides around 65% transmittance over an intended wavelength range. This designed SQWR will find potential application in the MSPI system used for different purposes to control polarization states of the incoming light wave. Furthermore, it will find application in optical metrological systems used over that range of frequencies where the quarter wave plate is a prerequisite, e.g., polarimetry of astronomical objects and optical isolators. All three materials show acceptable optical, thermal and mechanical properties over the intended frequency range of the spectrum.

References

1. S. Pancharatnam, Achromatic combinations of birefringent plates. Part II. An achromatic quarter wave plate, in *Proceedings of the Indian Academy of Sciences*, vol. XLI(4), Sec. A (1955), p. 137
2. C.M. McIntyre, S.E. Harris, Achromatic wave plates for the visible spectrum. J. Opt. Soc. Am. **58**(12), 1575–1580 (1968)
3. N. Mukhopadhyay, A. Saha, K. Bhattacharya, Study on superachromatism of a quarter-wave retarder for the visible range of the spectrum. J. Opt. Technol. **87**, 638–641 (2020)
4. P. Hariharan, Achromatic retarder using quartz and mica. Meas. Sci. Technol. **6**, 1078 (1995)
5. M.C. Gardner, P.J. Rogers, M.F. Wilde, T. Cook, A. Shipton, Challenges and solutions for high performance SWIR lens design, in *Proceedings of the SPIE 9987, Electro-Optical and Infrared Systems: Technology and Applications*, 99870C (October 21, 2016), p. XIII. https://doi.org/10.1117/12.2241089
6. J.N. Vizgaitis, B.F. Andresen, P.L. Marasco, J.S. Sanghera, S.P. Snyder, *Proceedings of the SPIE 9822, Advanced Optics for Defense Applications: UV through LWIR*. 982201 (July 18, 2016). https://doi.org/10.1117/12.2244400
7. J.H. Hough, Polarimetry Techniques at Optical and Infrared Wavelengths. *Astronomical Polarimetry: Current Status and Future Directions ASP Conference Series*, vol. 343 (2005)
8. F.F. Sabins, *Remote Sensing: Principles and Applications*. 3rd ed., Technology & Engineering (Waveland Press, June 11, 2007), p. 494
9. T. Bauer, H. Thome, T. Eisenhammer, Bandpass filter arrays patterned by photolithography for multispectral remote sensing, in *Proceeding of the SPIE 9241, Sensors, Systems, and Next-Generation Satellites XVIII, 92411K* (October 7, 2014). https://doi.org/10.1117/12.2069596
10. W.B. Sparksa, J. Hough, T.A. Germer, F. Chen, S. DasSarma, P. DasSarma, F.T. Robb, N. Manset, L. Kolokolova, N. Reid, D. Macchetto, W. Martin, Detection of circular polarization in light scattered from photosynthetic microbes, in *Proceedings of the National Academy of Sciences of the United States of America*, vol. 106, no. 19 (2009), pp. 7816–7821. https://doi.org/10.1073/pnas.0810215106
11. Z. Huang, A. McWilliams, H. Lui, D.L. Mclean, S. Lam, H. Zeng, Near-infrared Raman spectroscopy for optical diagnosis of lung cancer. Int. J. Cancer **107**, 1047–1052 (2003)
12. A.V. Samoylov, V.S. Samoylov, A.P. Vidmachenko, A.V. Perekhod, Achromatic and super-achromatic zero order waveplates. J. Quant. Spectrosc. Radiat. Transfer **88**, 319–325 (2004)
13. J. Ma, J.S. Wang, C. Denker, H.M. Wang, Optical design of multilayer achromatic waveplate by simulated annealing algorithm. Chin. J. Astron. Astrophys. **8**, 349–361 (2008)

14. N. Mukhopadhyay, A. Saha, K. Bhattacharya, Super-achromatic quarter-wave phase retarder for visible, near infrared and short wave infrared region applications. Opt. Spectrosc. **128**(8), 1199–1204 (2019)
15. N. Mukhopadhyay, S. Mandal, K. Bhattacharya, A. Saha, Design of a superachromatic quarter-wave retarder for near-infrared region using flower pollination algorithm. Opt. Eng. **58**(9), 095101 (2019)
16. S. De, N. Mukhopadhyay, An achromatic quarter-wave and half wave phase retarder combination operating in near-infrared (NIR) region. Int. J. Eng. Adv. Technol. (IJEAT) **9**(1), 6029–6031(2019)
17. N. Passily, K. Ventola, P. Karvinen, J. Turunen, J. Tervo, Achromatic phase retardation by subwavelength gratings in total internal reflection.J. Opt. A Pure Appl. Opt. **10**, 015001 (6pp) (2008)
18. H. Kikuta, Y. Ohira, K. Iwata, Achromatic quarter-waveplates using the dispersion of form birefringence. Appl. Opt. **36**(7), 1566–1572 (1997)
19. G. Kang, Q. Tan, X. Wang, G. Jin, Achromatic phase retarder applied to MWIR & LWIR dual-band. Opt. Express **18**(2), 1695–1703 (2010)
20. G.G. Kang, Q.F. Tan, G.F. Jin, Optimal design of an achromatic angle-insensitive phase retarder used in MWIR imaging polarimetry. Chin. Phys. Lett. **26**(074218), 1–4 (2009)
21. L. Na, K. Wei-Jin, X. Feng, Y. Mao-Jin, Broadband achromatic phase retarder based on metalmultilayer dielectric grating. Chin. Phys. B **27**(5), 054202 (2018)
22. X. Nie, T.X. Wu, Y.-Q. Lu, Y.-H. Wu, X. Liang, S.T. Wu, Dual-frequency addressed infrared liquid crystal phase modulators with submillisecond response time. Mol. Cryst. Liq. Cryst. **454**(1), 123/[525]–133/[535] (2006). https://doi.org/10.1080/15421400600654256
23. M.J. Abuleil, I. Abdulhalim, Tunable achromatic liquid crystal waveplates. Opt. Lett. **39**(19), 5487–5490 (2014)
24. O. Aharon, I. Abdulhalim, Liquid crystal wavelength independent continuous polarization rotator. Opt. Eng. **49**(3), 034002 (2010)
25. M.C. Nuss, J. Orenstein, *Terahertz Time-Domain Spectroscopy. Millimeter and Submillimeter Wave Spectroscopy of Solids* (Springer, 1998), pp. 7–50
26. M. Nagel, P. Haring Bolivar, M. Brucherseifer, H. Kurz, Integrated THz technology for label free genetic diagnostics. Appl. Phys. Lett. **80**, 154–156 (2002)
27. R.M. Woodward et al., Terahertz pulse imaging in reflection geometry of human skin cancer and skin tissue. Phys. Med. Biol. **47**, 3853–3863 (2002)
28. M. Yamashita, K. Kawase, C. Otani, K. Nikaw, M. Tonouchi, *in Conference on Optical Terahertz Science and Technology*, doc ID: TuC4 (Florida, USA, 2005)
29. B.M. Fischer et al., Terahertz time-domain spectroscopy and imaging of artificial RNA. Opt. Express **13**, 5205–5215 (2005)
30. P.F. Taday, I.V. Bradley, D.D. Arnone, M. Pepper, Using terahertz pulse spectroscopy to study the crystalline structure of a drug: a case study of the polymorphs of ranitidine hydrochloride. J. Pharm. Sci. **92**, 831–838 (2003)
31. M. Nagai, H. Yada, T. Arikawa, K. Tanaka, Terahertz time-domain attenuated total reflection spectroscopy in water and biological solution. Int. J. Infrared Milli. Waves **27**, 505–515 (2006)
32. K. Kawase, Y. Ogawa, Y. Watanabe, H. Inoue, Non-destructive terahertz imaging of illicit drugs using spectral fingerprints. Opt. Express **11**, 2549–2554 (2003)
33. H. Zhong, N. Karpowicz, X.-C. Zhang, Terahertz emission profile from laser-induced air plasma. Appl. Phys. Lett. **88**, 261103 (2006)
34. M.K. Choi, A. Bettermann, D.W. Van Der Weide, Potential for detection of explosive and biological hazards with electronic terahertz systems. Philos. Trans. R. Soc. Lond. A Math. Phys. Eng. Sci. **2004**(362), 337–349 (1815)
35. M.R. Leahy-Hoppa, M.J. Fitch, X. Zheng, L.M. Hayden, R. Osiander, Wideband terahertz spectroscopy of explosives. Chem. Phys. Lett. **434**(4), 227–230 (2007)
36. S. Nashima, O. Morikawa, K. Takata, M. Hangyo, Temperature dependence of optical and electronic properties of moderately doped silicon at terahertz frequencies. J. Appl. Phys. **90**, 837–842 (2001)

37. D.M. Mitttleman, J. Cunningham, M.C. Nuss, M. Geva, Noncontact semiconductor wafer characterization with the Hall effect. Appl. Phys. Lett. **71**, 16–18 (1997)
38. H. Akaike et al., Demonstration of a 129 GHz single-flux-quantum shift register circuit based on a 10kA cm^{-2} Nb process. Supercond. Sci. Technol. **19**, S320–S324 (2006)
39. A. Hirata, et al., 120-GHz-band millimeter-wave photonic wireless link for 10-Gb/s data transmission. IEEE Trans. Microwave Theory Technol. **54**, 1937–1944 (2006)
40. J.-B. Masson, G. Gallot, Terahertz achromatic quarter-wave plate. Opt. Lett. **31**, 265–267 (2006)
41. Z. Chen, Y. Gong, H. Dong, T. Notake, H. Minamide, Terahertz achromatic quarter wave plate: design, fabrication, and characterization. Opt. Commun. **311**, 1–5 (2013)
42. B. Zhang, Y. Gong, Achromatic terahertz quarter waveplate based on silicon grating. Opt. Express **23**, 14897–14902 (2015)
43. C. Hsieh, C. Yang, F. Shih, R. Pan, C. Pan, Liquid-crystal-based magnetically tunable terahertz achromatic quarter-wave plate. Opt. Express **27**, 9933–9940 (2019)
44. D.J. Diner, et al., *UV-SWIR Achromatic Quarter Wave Retarder for the Multiangle Spectropolarimetric Imager (MSPI).* College of Optical Sciences, University of Arizona, Tucson, AZ 85721 USA
45. N. Mukhopadhyay, A. Saha, K. Bhattacharya, *Multi-crystal Achromatic Quarter Wave Retarder for the Air-Multiangle Spectropolarimetric Imager (AirMSPI) in SWIR region* (IEMENTech2018, IEEE Digital Explore, 2018), pp. 1–3. https://doi.org/10.1109/IEMENTECH.2018.8465355
46. J.M. Bennet, Polarizers, in *Handbook of Optics*, ed. by M. Bass, vol. 2 (McGraw-Hill, New York, 1995), Chap. 3, table 8, pp. 3.46–3.49
47. X.-S. Yang, Flower pollination algorithm for global optimization. Lect. Notes Comput. Sci. **7445**, 240–249 (2012)
48. G. Ghosh, Dispersion-equation coefficients for the refractive index and birefringence of calcite and quartz crystals. Opt. Commun. **163**, 95–102 (1999)
49. D.E. McCarthy, Transmittance of optical materials from 0.17 μ to 3.0 μ. Appl. Opt. **6**(11), 1896–1898 (1967). https://doi.org/10.1364/AO.6.001896
50. J.L. Tosi, K.M. Khajurivala, *Janos Technology LLC: Photonics Handbook, Common Infrared Optical Materials and Coatings: A Guide to Properties, Performance and Applications*
51. N. Zhai, C. Li, B. Xu, L. Bai, J. Yao, G. Zhang, Z. Hu, Y. Wu, Temperature-dependent sellmeier equations of IR nonlinear optical crystal $BaGa_4Se_7$. Crystals **62**(7) (2017)

On the Photoemission from Quantum Confined Heavily Doped THz Materials

R. Paul, M. Mitra, S. Chakrabarti, and K. P. Ghatak

1 Introduction

The importance of Einstein's photoemission (EP) is already well-known since the discovery of Einstein's photoelectric effect and finds extensive applications in studying the physical properties of THzMs [1–10]. The THzMs are being widely studied in the literature [11–20]. The EP affects all the optical properties in a fundamental way since at terahertz frequency the band structure of THzMs changes radically. The EP is studied for all types of quantum-confined HD THzMs as stated in the abstract.

R. Paul
Department of Computer Science and Engineering, University of Engineering and Management, Kolkata 700156, India

M. Mitra
Department of Electronics and Telecommunication Engineering, Indian Institute of Engineering Science and Technology, Shibpur, Botanic Garden, Howrah, West Bengal 711103, India

S. Chakrabarti
University Of Engineering & Management, New Town, University Area, Plot No. III, B/5, New Town Rd, Action Area III, Kolkata 700160, India

Institute of Engineering and Management, Gurukul, Sikar Road NH 11, Near Udaipuria Mod, Kolkata and Jaipur, Rajasthan 303807, India

K. P. Ghatak (✉)
Department of Basic Science and Humanities, Institute of Engineering and Management, 1, Management House, Salt Lake, Sector-V, Kolkata 700091, India
e-mail: kamakhyaghatakcu@gmail.com

A. Saha et al. (eds.), *Optical to Terahertz Engineering*, Progress in Optical Science and Photonics 23, https://doi.org/10.1007/978-981-99-0228-6_3

2 Method

The EP from QWs, NWs, and QDs of HD THzMs can, respectively, be written as [21–32]

$$J_{2DL} = \frac{\alpha_0 g_v e}{\pi \hbar^2 d_z}\left(\frac{m_c}{2}\right)^{1/2} \text{Real Part of} \sum_{n_{z71\max}}^{n_{z71\min}} \left[\frac{\sqrt{T_1(E_{n_{z71}}, \eta_g, \lambda)}}{T_1'(E_{n_{z71}}, \eta_g, \lambda)}\right] [\phi_{71}(E_{F2DL}, n_{z71}) + \phi_{72}(E_{F2DL}, n_{z71})] \quad (1)$$

where $n_{z71\min} \geq \left(\frac{d_z}{\pi}\right)\left(\frac{\sqrt{2m_c}}{\hbar}\right)[T_1(W - h\upsilon, \eta_g, \lambda)]^{1/2}$

$$I_L = \frac{\alpha_0 e g_v k_B T}{\pi \hbar} \sum_{n_{7i}=1}^{n_{x7i\max}} \sum_{n_{7i}=1}^{n_{z7i\max}} F_0(\eta_i) \quad (2)$$

where $n_{7i} = \frac{E_{F1DL} - (E_{7i} + W - h\upsilon)}{k_B T}$ and

$$J_{0DL} = \frac{(\alpha_0 e g_v)}{d_x d_y d_z}\left(\frac{m_c}{2}\right)^{-1/2} \text{Real Part of} \sum_{n_{x71}=1}^{n_{x71\max}} \sum_{n_{y71}=1}^{n_{y71\max}} \sum_{n_z 71\min}^{n_{z71\max}} \left[\frac{\sqrt{T_1(E_{n_{z71}}, \eta_g, \lambda)}}{T_1'(E_{n_{z71}}, \eta_g, \lambda)}\right] F_{-1}(\eta_{71OD}) \quad (3)$$

where the notations are defined above.

3 Results and Discussions

Using the appropriate equations, we have plotted the normalized EP (ϕ_1) as functions of d_z, , normalized dimension-dependent carrier statistics (ϕ_2), normalized wavelength (ϕ_3), and normalized intensity (ϕ_4) in Figs. 1, 2, 3, 4, 5, 6, 7, 8, 9, 10, 11, 12, 13, 14, 15, 16, 17, 18, 19, 20, 21, 22, 23 and 24, respectively. From the said figures, the following features follow:

1. From Figs. 1, 5, 9, 13, 17, and 21, we note that the ϕ_1 decreases with increasing d_z in oscillatory ways for QWs, NWs, and QDs of HD THzMs, respectively.
2. From Figs. 2, 6, 10, 14 18, and 22, we observe that ϕ_1 increases with increasing ϕ_2 in oscillatory ways for QWs, NWs, and QDs of HD THzMs, respectively.
3. From Figs. 4, 8, 12, 16, 20, and 24, we note that ϕ_1 decreases with increasing ϕ_3 in monotonic fashion for QWs, NWs, and QDs of HD THzMs, respectively.
4. From Figs. 3, 7, 11, 15, 19, and 23, we note that ϕ_1 increases with decreasing ϕ_4 in monotic fashion for QWs, NWs, and QDs of HD THzMs, respectively.

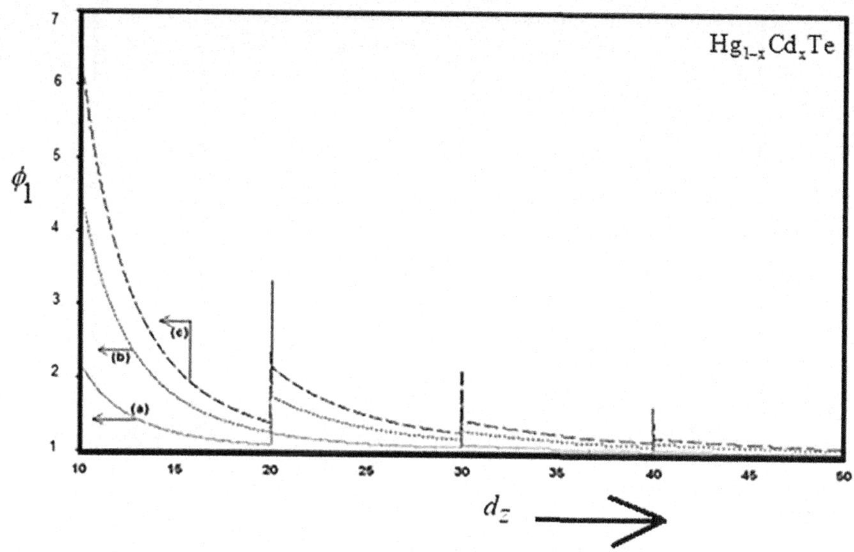

Fig. 1 Plot of ϕ_1 from QWs of the indicated HD material versus d_z for **a** 3-band Kane model **b** 2-band Kane model **c** parabolic band model, respectively

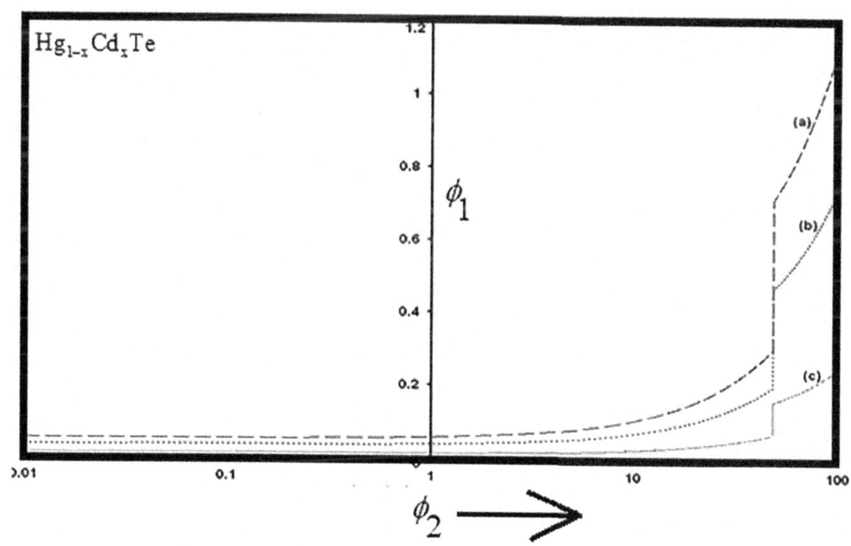

Fig. 2 Plot of ϕ_1 from QWs of the indicated HD material versus ϕ_2 for all cases in Fig. 1

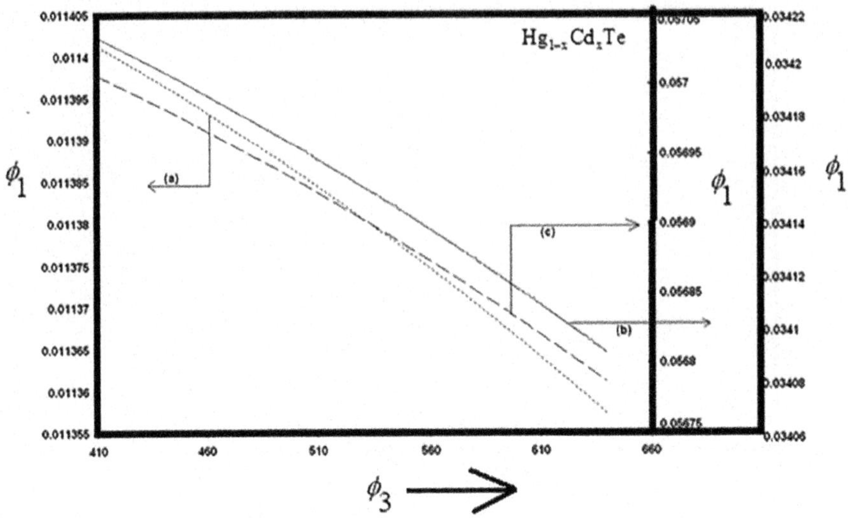

Fig. 3 Plot of ϕ_1 from QWs of the indicated HD material versus ϕ_3 for all cases in Fig. 1

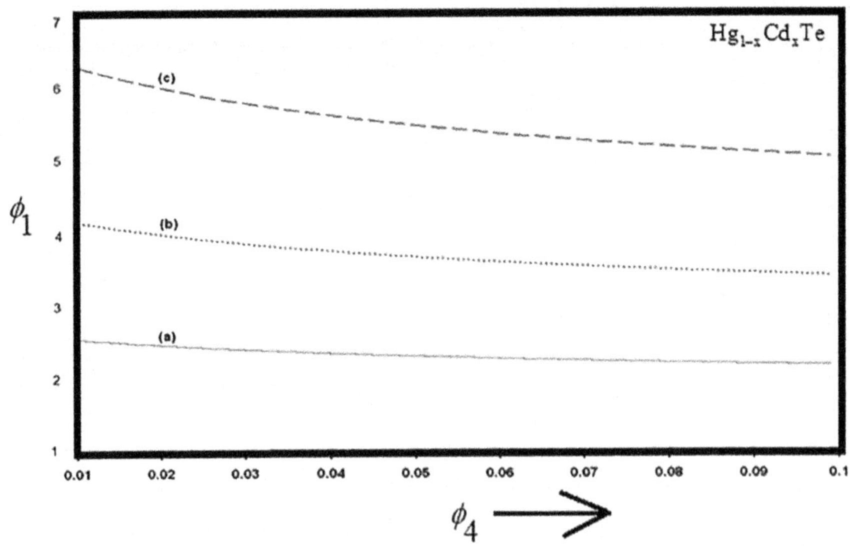

Fig. 4 Plot of ϕ_1 from QWs of the indicated HD material versus ϕ_4 for all cases in Fig. 1

4 Conclusion

Einstein's photoemission has been investigated in QWs, NWs, and QDs of heavily doped THzMs taking HD $Hg_{1-x}Cd_xTe$ and HD $n-In_{1-x}Ga_xAs_yP_{1-y}$ as examples. We note that photoemission decreases with film thickness and increases with

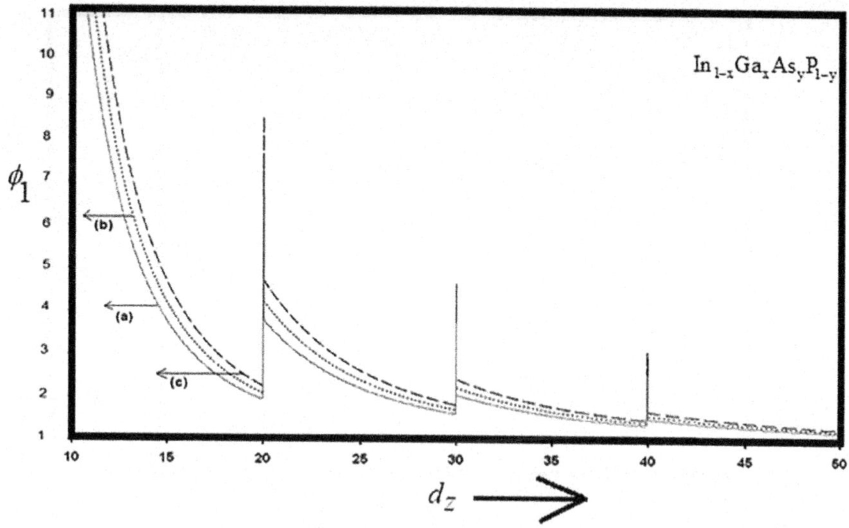

Fig. 5 Plot of ϕ_1 from QWs of the indicated HD material versus d_z for all the cases in Fig. 1

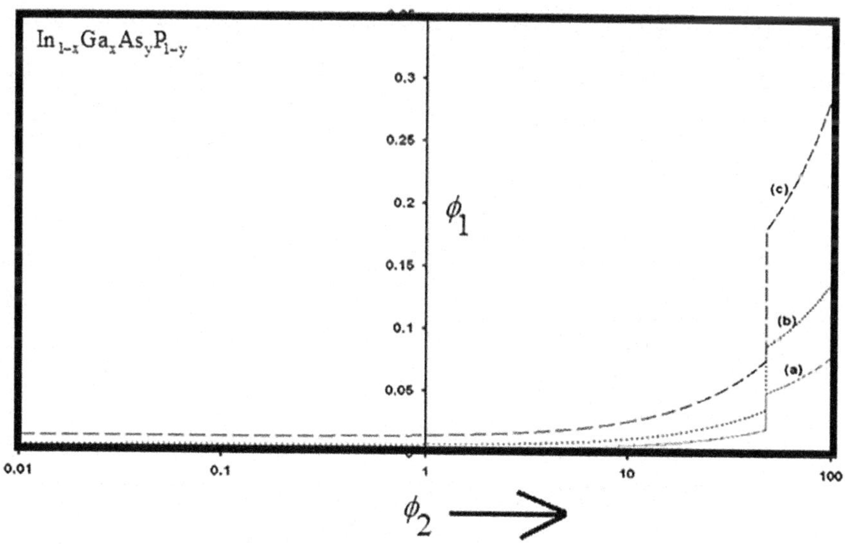

Fig. 6 Plot of ϕ_1 from QWs of the indicated HD material versus ϕ_2 for all cases in Fig. 1

electron degeneracy in oscillatory ways. Besides the photoemission decreases with wavelength and intensity, respectively, for the entire quantum confined cases. The quantum signature of Einstein's Photoemission for all quantum-confined cases can easily the assessed from the figures.

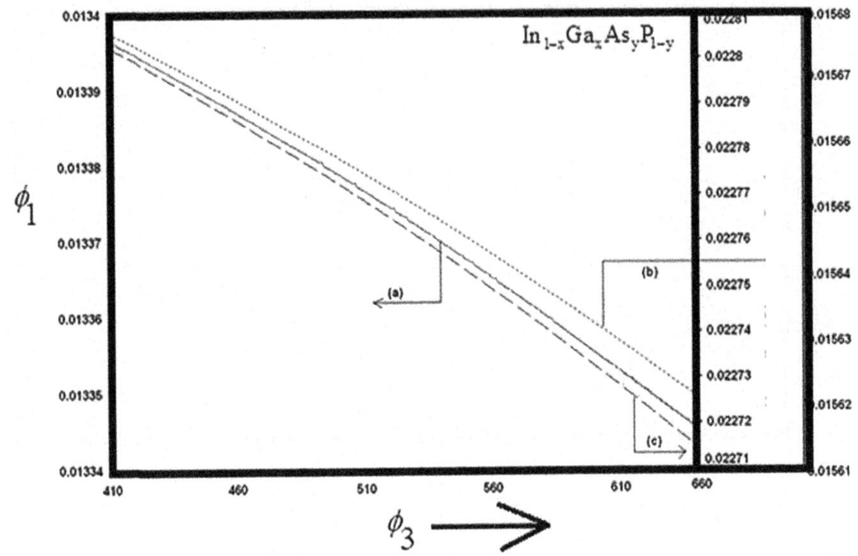

Fig. 7 Plot of ϕ_1 from QWs of the indicated HD material versus ϕ_3 for all cases in Fig. 1

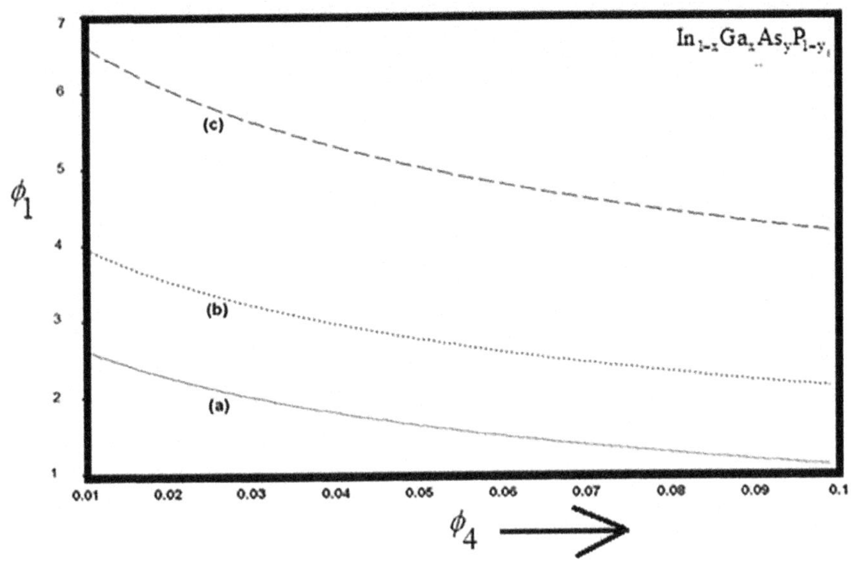

Fig. 8 Plot of ϕ_1 from QWs of the indicated HD material versus ϕ_4 for all cases in Fig. 1

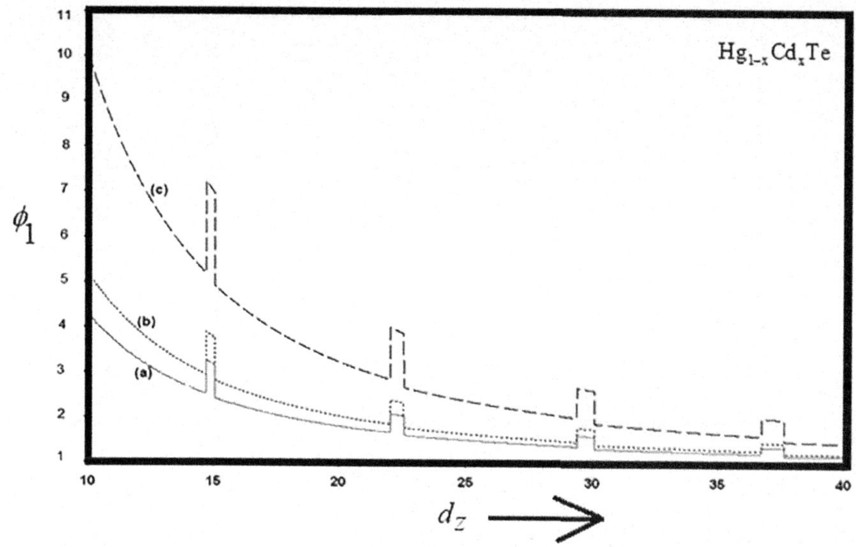

Fig. 9 Plot of ϕ_1 from NWs of the indicated HD material versus d_z for all cases in Fig. 1

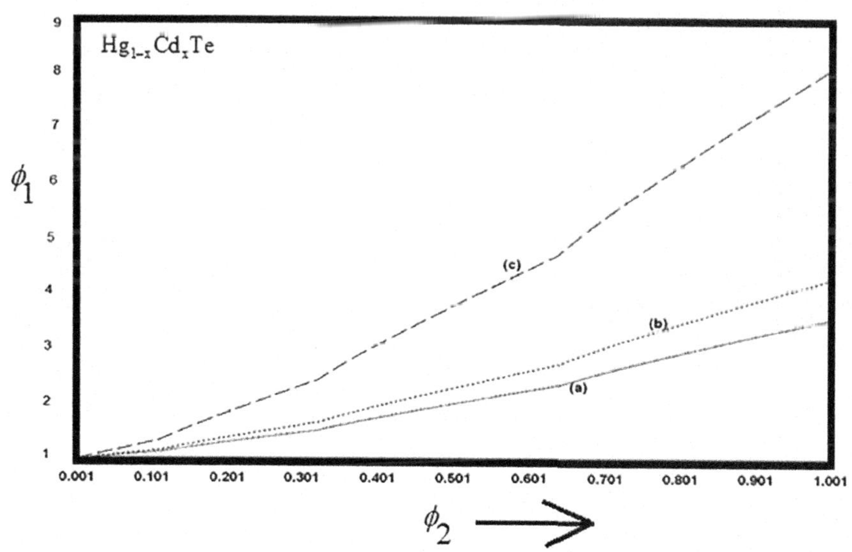

Fig. 10 Plot of ϕ_1 from NWs of the indicated HD material versus ϕ_2 for all cases in Fig. 1

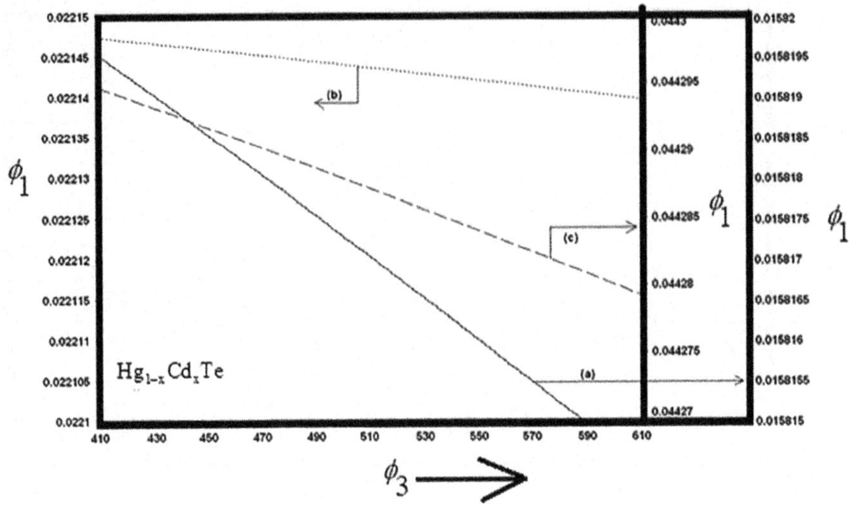

Fig. 11 Plot of ϕ_1 from NWs of the indicated HD material versus ϕ_3 for all cases in Fig. 1

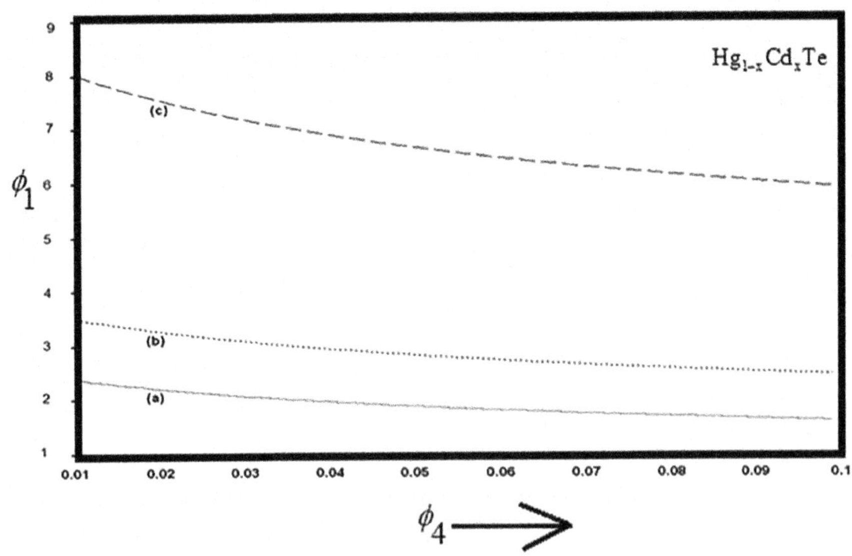

Fig. 12 Plot of ϕ_1 from NWs of the indicated HD material versus ϕ_4 for all cases in Fig. 1

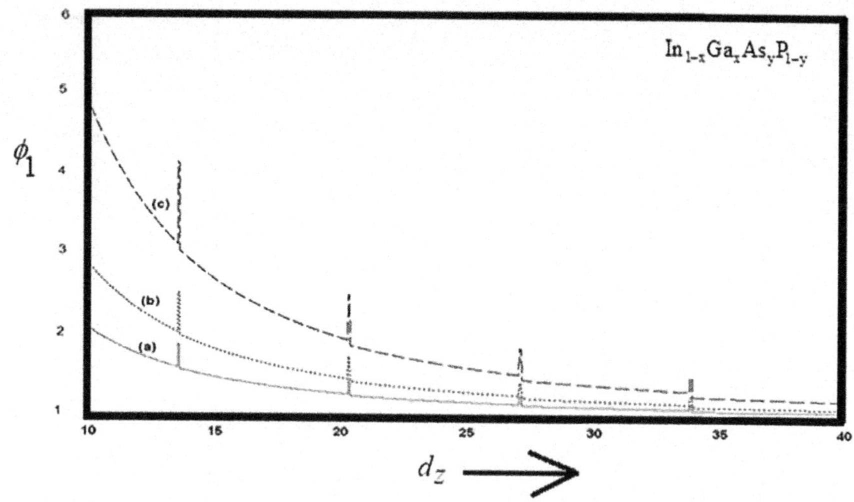

Fig. 13 Plot of ϕ_1 from NWs of the indicated HD material versus d_z for all cases in Fig. 1

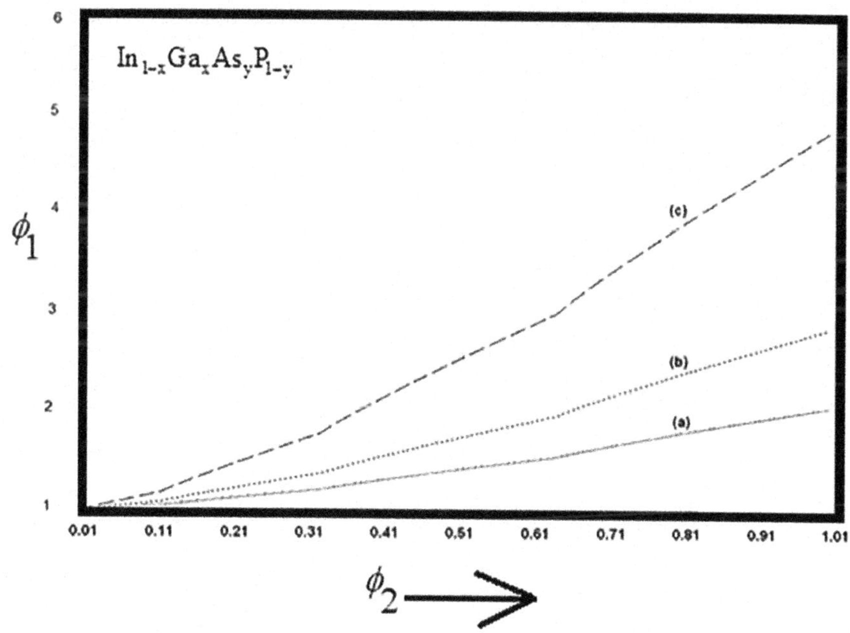

Fig. 14 Plot of ϕ_1 from NWs of the indicated HD material versus ϕ_2 for all cases in Fig. 1

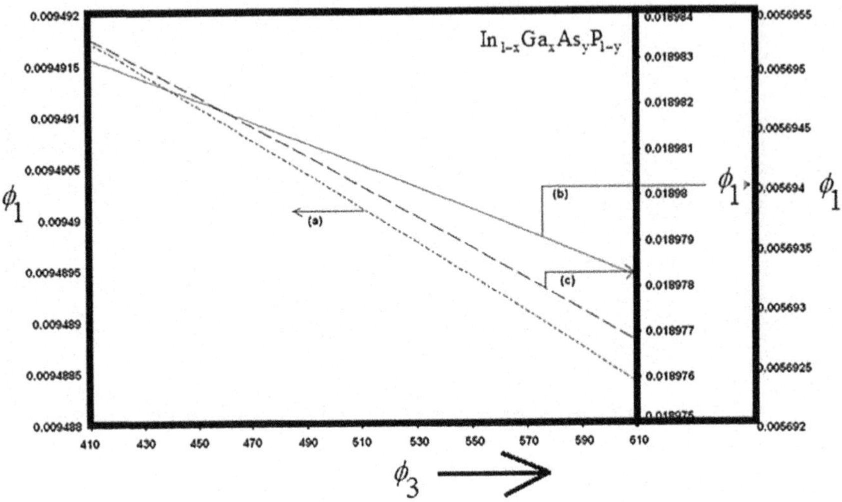

Fig. 15 Plot of ϕ_1 from NWs of the indicated HD material versus ϕ_3 for all cases in Fig. 1

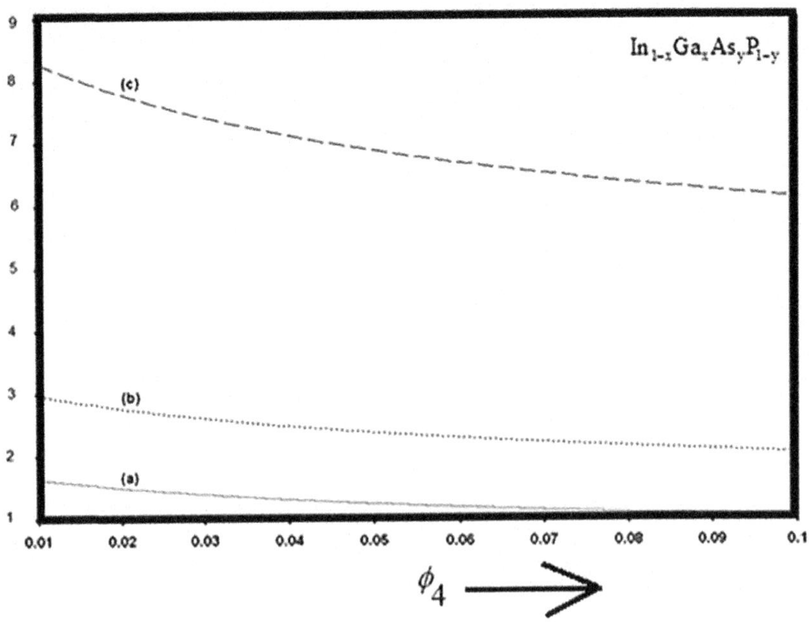

Fig. 16 Plot of ϕ_1 from NWs of the indicated HD material versus ϕ_4 for all cases in Fig. 1

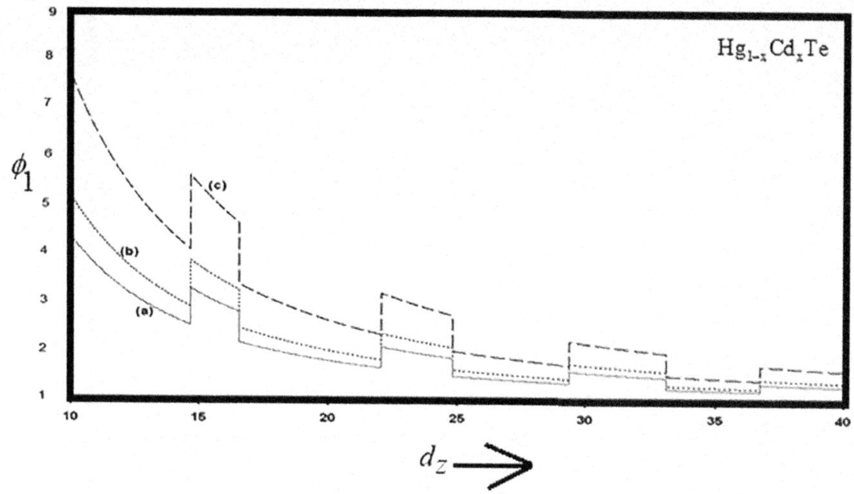

Fig. 17 Plot of ϕ_1 from QDs of the indicated HD material versus d_z for all cases in Fig. 1

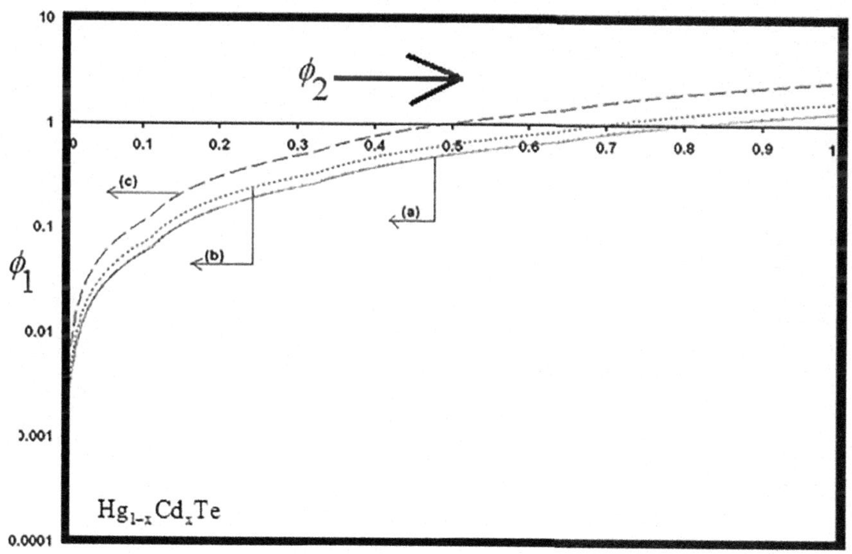

Fig. 18 Plot of ϕ_1 from QDs of the indicated HD material versus ϕ_2 for all cases in Fig. 1

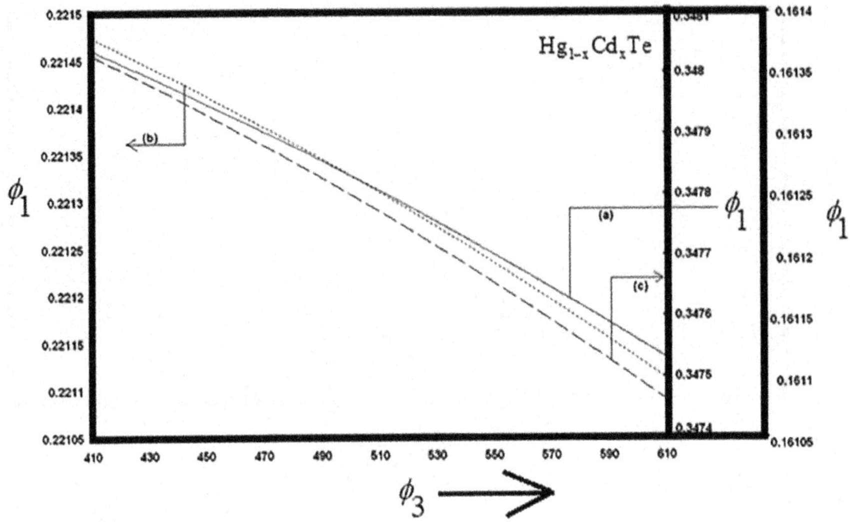

Fig. 19 Plot of ϕ_1 from QDs of the indicated HD material versus ϕ_3 for all cases in Fig. 1

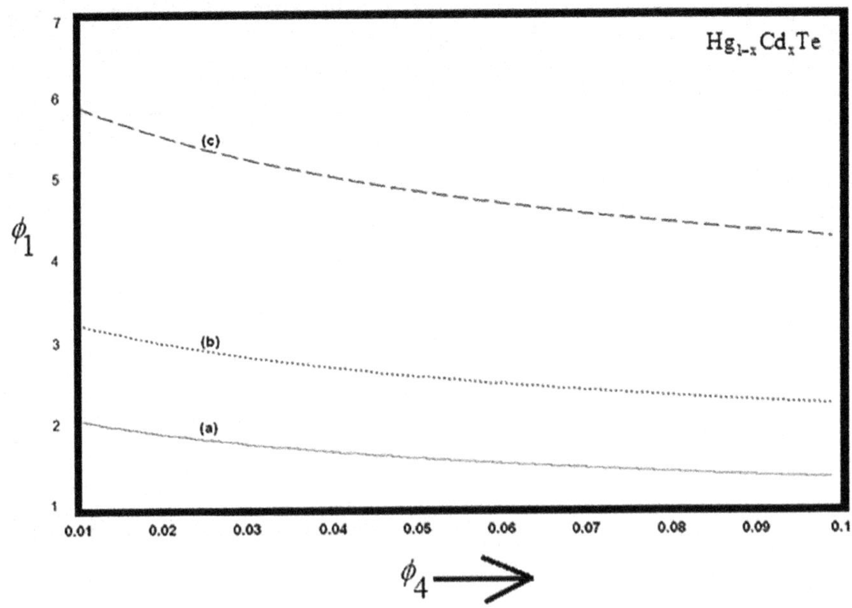

Fig. 20 Plot of ϕ_1 from QDs of the indicated HD material versus ϕ_4 for all cases in Fig. 1

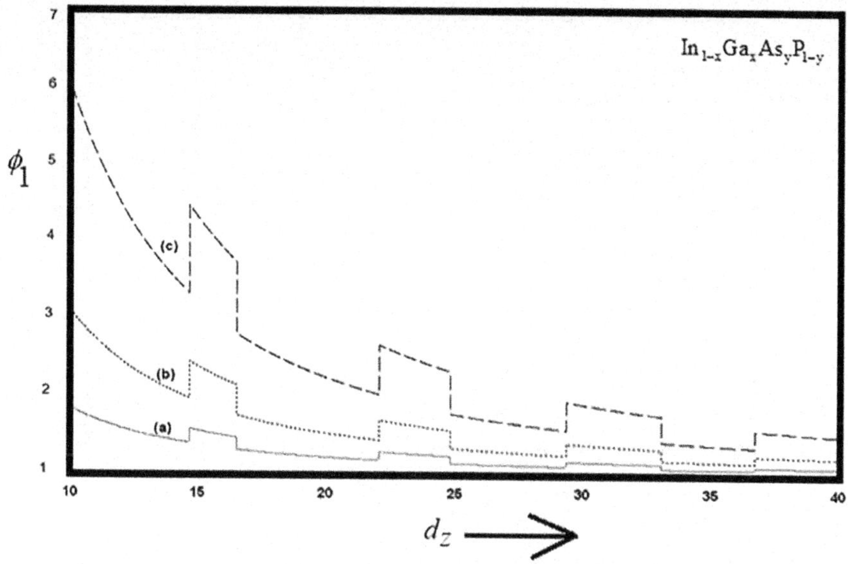

Fig. 21 Plot of ϕ_1 from QDs of the indicated HD material versus d_z for all cases in Fig. 1

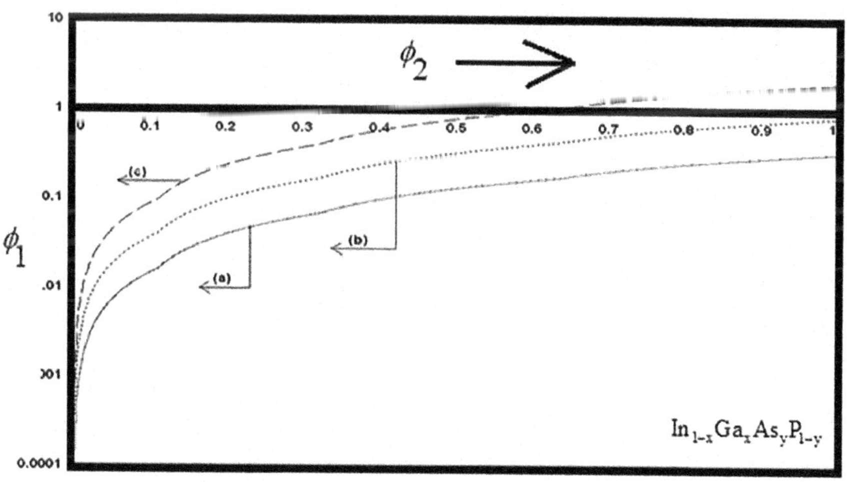

Fig. 22 Plot of ϕ_1 from QDs of the indicated HD material versus ϕ_2 for all cases in Fig. 1

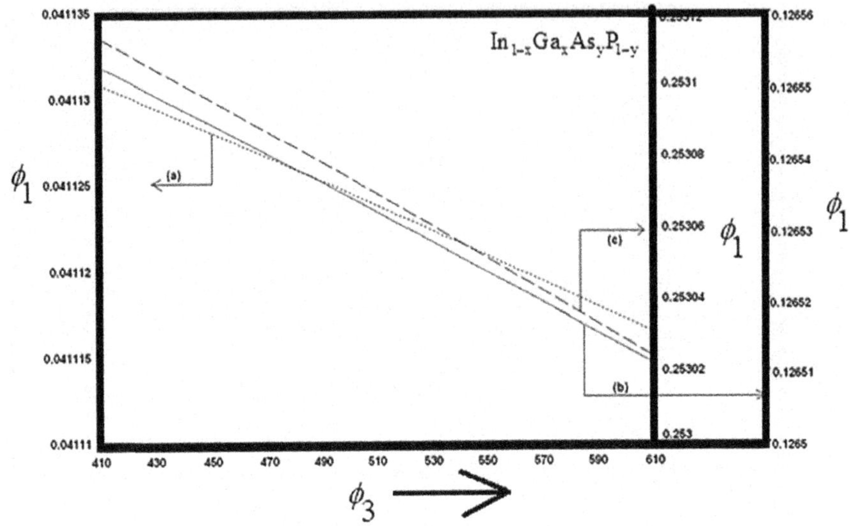

Fig. 23 Plot of ϕ_1 from QDs of the indicated HD material versus ϕ_3 for all cases in Fig. 1

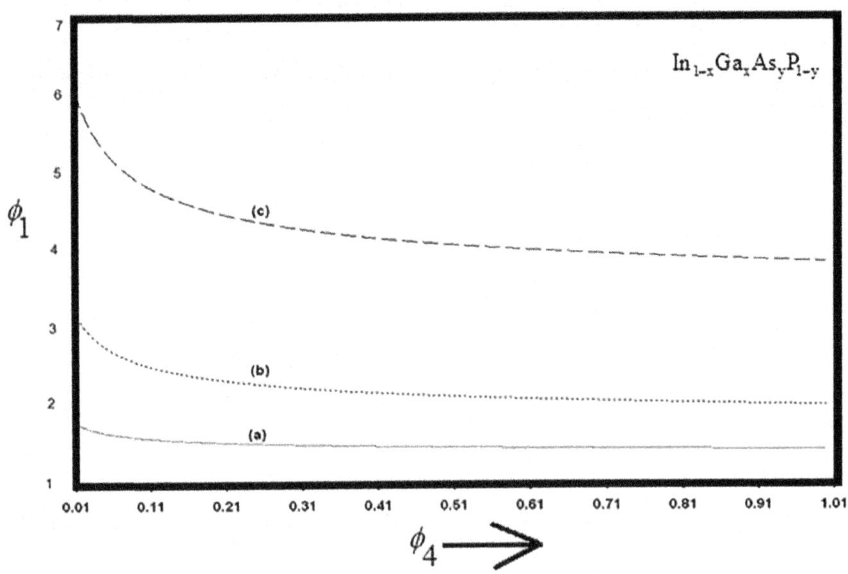

Fig. 24 Plot of ϕ_1 from QDs of the indicated HD material versus ϕ_4 for all cases in Fig. 1

Acknowledgements The authors are grateful to Prof. Dr. S. Chakrabarti, the Hon'ble President of IEM UEM Group, Kolkata, India for his constant inspiration and discussion in spite of very busy schedule.

References

1. M. Cardona, L. Ley, *Photoemission in Solids 1 and 2, Topics in Applied Physics*, vols. 26, 27 (Springer, Germany, 1978)
2. S. Hüfner, *Photoelectron Spectroscopy* (Springer, Germany, 2003)
3. S. Hüfner (ed.), *Very High Resolution Photoelectron Spectroscopy*. Lecture Notes in Physics, vol. 715 (Springer, Germany, 2007)
4. D.W. Lynch, C.G. Olson, *Photoemission Studies of High-Temperature Superconductors* (Cambridge University Press, UK, 1999)
5. *Photoemission and the Electronic Properties of Surfaces*, ed. by B. Feuerbacher, B. Fitton, R.F. Willis (Wiley, New York, 1978)
6. W. Schattke, M.A.V. Hove, *Solid-State Photoemission and Related Methods: Theory and Experiment* (Wiley, USA, 2003)
7. V.V. Afanas'ev, *Internal Photoemission Spectroscopy: Principles and Applications* (Elsevier, North Holland, 2010)
8. D.J. Lockwood, *Light Emission in Silicon in silicon based materials and devices*, ed. by H.S. Nalwa, vol. 2 (Academic Press, San Diego, USA, 2001)
9. K.P. Ghatak, D. De, S. Bhattacharya, *Photoemission from Optoelectronic Materials and their Nanostructures*. Springer Series in Nanostructure Science and Technology (Springer, 2009)
10. T.N. Sen, K.P. Ghatak, Einstein's 1D photo emission and the Von Klitzing constant. J. Nanosci. Nanotechnol. **16**, 1229–1230 (2016)
11. S. Debbarma, K.P. Ghatak, Einstein's photoemission from quantum confined superlattices. J. Nanosci. Nanotechnol. **16**, 1095–1124 (2016)
12. K. Sarkar, M. Chakraborty, S. Chakravarti, B. Chatterjee, K.P. Ghatak, Quantum confined heavily doped optoelectronic materials and the Einstein's photoemission. J. Nanoeng. Nanomanuf. **5**, 43–63 (2015)
13. S.M. Adhikari, K.P. Ghatak, The photoemission from quantum wires of heavily doped nonparabolic materials. J. Adv. Phys. **2**, 130–142 (2013)
14. D. De, A. Kumar, S.M. Adhikari, S. Pahari, N. Islam, P. Banerjee, S.K. Biswas, Influence of quantum confinement on the photoemission from superlattices of optoelectronic materials. Superlattices Microstruct. **47**, 377–410 (2010)
15. A. Kumar, S. Chowdhury, S.M. Adhikari, S. Ghosh, M. Mitra, D. De, A. Sharma, S. Bhattacharya, A. Dey, K.P. Ghatak, The photoemission from quantum wells, wires and carbon nanotubes: simplified theory and relative comparison. J. Comput. Theor. Nanosci. **7**, 115–145 (2010)
16. S. Pahari, S. Bhattacharya, S. Roy, A. Saha, D. De, K.P. Ghatak, Simple theoretical analysis of the Einstein's photoemission from quantum confined superlattices. Superlattices Microstruct. **46**, 760–796 (2009)
17. K.P. Ghatak, J. Mukhopadhyay, J.P. Banerjee, On the photo emission from A3IIB2V nanostructures, in *SPIE Proceedings Series* (2002), pp. 1296–1299
18. A.P. Ghatak, P.K. Bose, G. Majumder, Influence of quantum confinement on the photoemission from nonlinear optical materials. MRS Online Proc. Libr. (OPL) **484** (1997)
19. K.P. Ghatak, B. Nag, G. Mazumder, On the photoemission from quantum confined strained III–V systems, MRS Online Proc. Libr. (OPL) **379** (1995)
20. K.P. Ghatak, B. De, The photoemission from superlattices of III–V semiconductors with graded interfaces under quantizing magnetic field. MRS Online Proc. Libr. (OPL) **299** (1994)

21. K.P. Ghatak, Photoemission from superlattices of nonlinear optical materials with graded structures under strong magnetic field, in *SPIE Proceedings Series, Nonlinear Optics III*, vol. 1626 (1992), pp. 115–125
22. K.P. Ghatak, A. Ghoshal, B. De, Effect of photon energy on the photoemission from nonlinear optical materials under different physical conditions, in *Optoelectronic Devices and Applications, SPIE Proceedings Series*, vol. 1338 (1990), pp. 111–133
23. K.P. Ghatak, M. Mondal, S.N. Biswas, On a simplified analysis of the photoemission from quaternary alloys. J. Appl. Phys. **68**, 3032–3034 (1990)
24. K.P. Ghatak, On the photoemission from 3-D quantum well boxes of nonlinear optical materials in the presence of crossed electric and magnetic fields, in *High Speed Phenomena in Photonic Materials and Optical Bistability, SPIE Proceedings Series*, vol. 1280 (1990), pp. 53–67
25. K.P. Ghatak, A. Ghoshal, S. Bhattacharyya, Influence of photon energy on the photoemission from nonlinear optical materials and devices under different physical conditions, in *Nonlinear Optical Materials and Devices for Photonic Switching, SPIE Proceedings Series*, vol. 1216 (1990), pp. 282–290
26. K.P. Ghatak, The influence of pho ton energy on the photoemission from nonlinear optical materials from ultrathin films, quantum well wires, quantum dots, in *Optoelectronic Devices and Applications, SPIE Proceedings Series, 10–11 July 1990, San Diego* (1990)
27. K.P. Ghatak, On the photoemission from ternary and quaternary alloys systems under strong magnetic quantization. MRS Online Proc. Libr. (OPL) **216** (1990)
28. K.P. Ghatak, A. Ghoshal, S.N. Biswas, On the photo-emission from ultrathin films of optical materials current developments in optical engineering and commercial optics, in *SPIE*, vol. 1168 (1989), pp. 229–242
29. M. Mondal, S. Banik, K.P. Ghatak, Theoretical analysis of the photoemission from quantum well wires of ternary chalcopyrite semiconductors. J. Low Temp. Phys. **74**, 423–433 (1989)
30. B. Mitra, K.P. Ghatak, On the photoemission from quantum well wires of bismuth. Phys. Scr. **40**, 776 (1989)
31. B. Mitra, A. Ghosal, K.P. Ghatak, On the photoemission from Bi in the presence of an arbitrarily oriented quantizing field. physica status solidi (b), **150**, K67-K72 (1988)
32. K.P. Ghatak, M. Mondal, On the photoemission from ultrathin films of n-Ge. Solid State Electron. **31**, 1561–1562 (1988)

Evolution of 6G and Terahertz Communication

Pia Sarkar and Arijit Saha

1 Introduction

There has been a phenomenal advancement of mobile communication and wireless networks over the last decade. The ever-growing demand of users leads to huge data traffic and bandwidth requirements. As we are gradually heading towards futuristic 6G communications, it will require huge data rates, possibly around 10 Gigabits per second (Gbps) to cater lots of resources, which will again require very high frequencies even more than 100 GHz. In fact for Terahertz (THz) communication, realization of faster data transmission rate of the order of Gbps and Terabits per second (Tbps) has introduced a new area of research. Data rate and capacity can be improved implementing Massive Multiple Input Multiple Output (MIMO) [2] and backhaul [3–5]. According to Institute of Electrical and Electronics Engineers (IEEE) standards of Terahertz Science and Technology, 300 GHz–10 THz is considered as Terahertz band. Frequency beyond 100 Gigahertz (GHz) is labelled as THz by RF engineers.

2 Background

As an enhancement of cellular networks, 5G technology can improve the performance in terms of bandwidth (between 100 and 1000 MHz, i.e., short wavelength between 1 and 10 mm), speed (gigabits per second), spectral efficiency, latency (less than 5 ms), and energy savings. More promising applications, high connectivity between

P. Sarkar (✉) · A. Saha
Department of Electronics and Communication Engineering, B.P. Poddar Institute of Management and Technology, Kolkata 700 052, India
e-mail: pia.sarkar@bppimt.ac.in

A. Saha et al. (eds.), *Optical to Terahertz Engineering*, Progress in Optical Science and Photonics 23, https://doi.org/10.1007/978-981-99-0228-6_4

multiple devices in agriculture and construction industries is accommodated. The requirements of 5G technology are:

Ultra-reliable low latency communications (uRLLC) [6]: To reduce latency and increase speed to manage the timing of machine and secured system.

Enhanced mobile broadband (xMBB) [7]: It involves capacity enhancement.

Massive machine type communications (mMTC) [8]: It provides connections to large no of devices that intermittently transmit small amount of traffic.

Multicarrier modulation i.e., Orthogonal Frequency Division Multiplexing (OFDM) [9] is introduced in Multiple Input Multiple Output to improve the channel estimation efficiency. Minimum Mean Square Error (MMSE) and the Low Rank Approximation algorithms [9] provide good performance.

In comparison with 5G technology, 6G wireless communication provides small wavelength, wide bandwidth, higher data rate, better spectral efficiency, lower latency of 0.1 ms, and enhanced reliability.

Using heterogeneous networks and large numbers of antennas, 6G networks provide better performance. THz wave can be applied in spectroscopy, sensing, and communications.

3 Significance of Terahertz Frequency Bands

THz communications have become very attractive since they offer advantages of both millimetre wave and light wave. Some prominent advantages of THz communication are discussed below.

1. *Wide bandwidth*: THz communication links offer higher bandwidth compared to millimetre wave links, higher transmission rate, stronger directionality of the beam and better reliability. Eavesdropping [10] can be reduced. The size of the antenna will be reduced which favours smaller satellite systems.
2. *Less harmful*: As the energy of THz waves do not harm living organisms, it can be safely applied in biomedical sector.
3. *Good resolution*: THz radiation can be introduced for tracking suspicious objects.
4. *Less scattering loss*: THz and millimetre waves experience much less scattering loss caused by particulates compared to optical wavelengths. In case of rain or heavy clouds millimetre band communications can be used as a backup to an optical link and in case of heavy particulates (smoke, dust) THz can be used for short distances.

4 Challenging Issues in Terahertz Frequency Bands

1. In terahertz communications beam direction control is a challenge due to shorter wavelength.
2. Above 1 THz significant absorption occurs in the atmosphere due to water vapour.
3. Multipath and Doppler effects can affect signal transmission. As channel characteristics change with time, proper Channel estimation technique is required to improve spectral efficiency.
4. To design THz device, sampling frequency should be high. However, at present low efficiency and relatively low power are available from currently available sources.

5 Previous Works

Piesiewicz et al. [11] proposed the concept of using higher bandwidth in the terahertz frequency band. Federici et al. [12] explored different scopes of sub-THz and THz frequency bands. Huang et al. [13] presented the combination of analog sub-arrays and digital beamformer in digital signal processing. Huang et al. [14] gave an overview of the modern development in THz communication and its application in Terabit radio systems. Song and Nagatsuma et al. [15] discussed on the difficulties in designing THz devices and advantages of THz frequency bands. Kurner et al. [16] presented THz channel modelling and standardization techniques. Akyildiz et al. [17] provided a survey on modulation techniques to increase the capacity. They also highlighted on channel modelling. Akyildiz et al. [18] gave an overview about the new developments in THz devices, communication and networking. Guo et al. [19] summarized the 6G-enabled massive Internet-of-Things (IoT). Ul Hasan et al. [20] defined channel allocation mechanism to enable growing number of subscribers reducing the overall interference.

Nagatsuma et al. [21, 22] and Mehdi et al. [23] presented the recent advancements in THz device. Han et al. [24] presented the appropriate system model of different operations taking into account the environmental parameters. They also analysed the impact of misalignment and multipath fading. Wu et al. [25] proposed a new calculation to merge channels. Nguyen et al. [26] analysed applications above 100 GHz and focused on wave propagation properties at 140 GHz. Boulogeorgos et al. [27] highlighted the channel and baseband challenges and technique of delivering optical network quality in the THz range. Headland et al. [28] discussed experimental findings of the beam control in the terahertz range. K. Sengupta et al. [29] focused on the opto-electronic system. Akyildiz et al. [30] focused on the implementation of MIMO antenna. Rappaport et al. [31] presented the benefits and difficulties of using higher frequency band in 6G. Elayan et al. [32] introduced THz generation methods and proposed channel models at THz band frequencies. Corre et al. [33] explored the unused frequency range for optimization. Haselmayr et al. [34] highlighted on nano technology.

Lopez et al. [35] presented the challenges and techniques of Wireless Energy Transfer (WET). Piran et al. [36] considered the requirements, applications, and challenges of 6G systems. Stoica et al. [37] discussed the state of the art of AI in 6G. Kalbande et al. [38] presented the vision to implement 6G networks in India. Ghafoor et al. [39] overviewed the applications of THZ frequency. Tekbiyik et al. [40] proposed the heterogeneous architecture. Zhang et al. [41] presented optoelectronic terahertz communications technology with high data rate of multicarrier modulation. Polese et al. [42] gave a flavour of challenges to implement THz frequency in mobile communication. Sarieddeen et al. [43] discussed on sharing of resources. Tripathi et al. [1] gave a flavour of using millimetre wave and THz bands for commercial communication. They discussed propagation characteristics, channel models, design, and applications of 6G. Sarieddeen et al. [10] discussed THz signal generation and channel estimation. They also presented Massive Multiple-Input Multiple-Output systems. Lemic et al. [44] overviewed nano technology in THz communication.

Chaccour et al. [45] highlighted the technique of turning THz communication challenges into sensing opportunities. Tataria et al. [46] and Kaushal [47] overviewed the vision, strength and weakness of 6G. Elmeadawy et al. [48], Tariqet et al. [49], Gui et al. [50], Matthaiou et al. [51] presented the state of art techniques in 6G. Bariah et al. [52], Chen et al. [53], Yuan et al. [54], Chen et al. [55] overviewed the development of 6G. Giordani et al. [56] presented the recent 6G advancements. Elayan et al. [32] highlighted on the current research trends in the THz range. Mughees et al. [57] overviewed 6G technology. Tripathi et al. [1] differentiated between millimetre wave and THz communications. Chaccour et al. [45] focused on joint THz communications and sensing and possible applications.

6 Propagation Characteristics

Some of the important propagation characteristics are discussed here.

Atmospheric attenuation [58]: Caused by water vapour and oxygen absorption. It limits the transmission distance of the THz wave. The variation of the atmospheric refractive index changes the path of propagation.

Molecular absorption [10]: When air molecules are excited some energy is lost due to internal vibration.

Beamforming [10] *and precoding*: Sufficient power can be directed towards a direction using Beamforming. In wideband hybrid beamforming schemes [59], analog OFDM-based algorithm with digital precoding technique is applied.

7 Technologies Related to Channel, Antenna and Access Techniques

7.1 Channel Measurement Methods [60]

Three types of channel measurements methods are there - Terahertz Time Domain Spectroscopy (THz-TDS), Sliding-correlator (SC) and Vector Network Analyser (VNA)-swept-frequency method.

7.2 Channel Modelling Methods [61]

Deterministic, Statistical and hybrid channel modelling can be implemented.

7.3 Multicarrier Techniques [62]

Orthogonal Frequency Division Multiplexing (OFDM) improves capacity. However it increases Peak-to-Average Power Ratio (PAPR).

7.4 Channel Coding and Decoding [10]

There is a trade-off between latency and length of the code-word. Latency should be low in THz communication. If the length of the Codes becomes short its reliability will be less. As IoT system requires short packets, and error-free transmission cannot be guaranteed easily. Zero-forcing and MMSE technique [9] can optimize the performance.

7.5 Massive MIMO Systems [63–67]

In Multiple input multiple output (MIMO) multiple antenna elements with spacing half of the wavelength of the transmitted signal can be integrated as shown in Fig. 1. Multicarrier modulation OFDM technique can be adopted in wireless communication applying Fast Fourier Transform (FFT) and Inverse Fast Fourier Transform (IFFT). In Fig. 2 the MIMO-OFDM system model is shown where MIMO applies orthogonal frequency division multiplexing (OFDM). If the same signal is transmitted from the transmitter to the receiver through different paths fading can be mitigated. Capacity can be increased if spatial multiplexing is used. The output of MIMO is

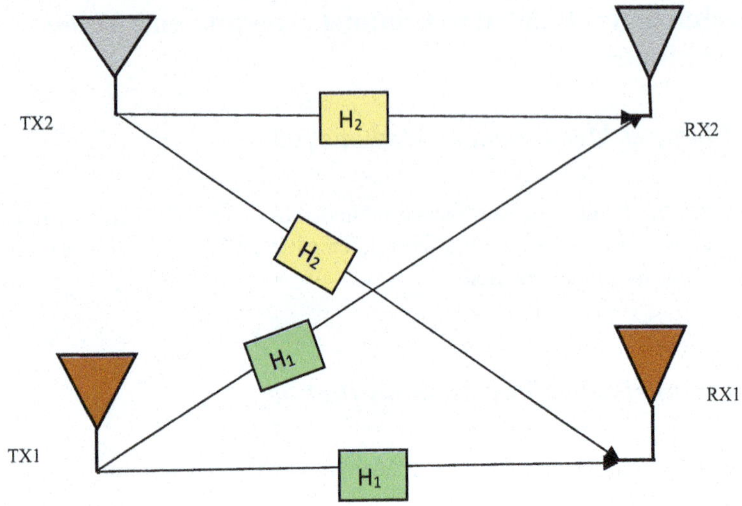

Fig. 1 MIMO channel model

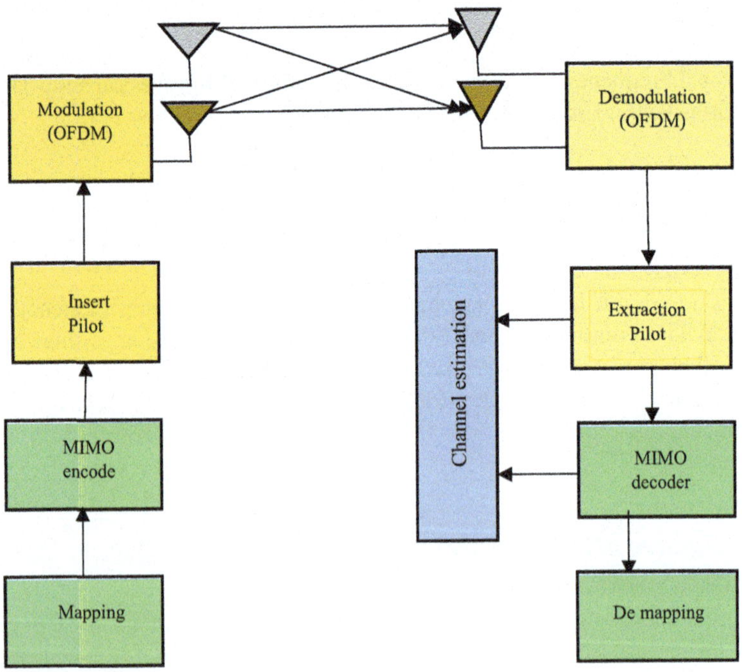

Fig. 2 MIMO-OFDM system model

$$Y = HX + N \tag{1}$$

where, H = Matrix of the channel, Y = Received signal, X = Transmitted signal and N = Noise signal.

In general, two transmitting antennas transmit to two receiving antennas. This system applies FFT and IFFT algorithm, and adds a pilot to estimate the channel. Before the OFDM symbol it includes a cyclic prefix of length more than the channel's maximum delay. This guard interval is used to reduce Inter Symbol Interference (ISI).

7.6 OAM-Based Systems [68]

Orbital Angular Momentum (OAM) system implements smaller antennas with different modes for optimization.

7.7 Non-orthogonal Multiple Access (NOMA) [69, 70]

When all users access the resources, this technique is applied to resist the spectral efficiency loss. In NOMA different powers P_1 and P_2 are transmitted from the Base station for different users U_1 and U_2. Let, $P_1 > P_2$. In Fig. 3, two users are there in cluster 1. User 1 (U_1) receives its own data. Data of User 1 is decoded by user 2 (U_2). Then, U_2 receives its own data after subtracting User 1's message. Both users can access the resources, whereas the user at a more distance place (in this case U_1) cannot interfere with the near user (in this case U_2). In Fig. 3 the NOMA-MIMO system is shown where resources are used in spatial dimension to improve the spectral efficiency and beam forming is applied to minimize the interference.

The challenging issues in NOMA-MIMO are Power and beamforming optimization and Successive Interference Cancellation (SIC)-Stability.

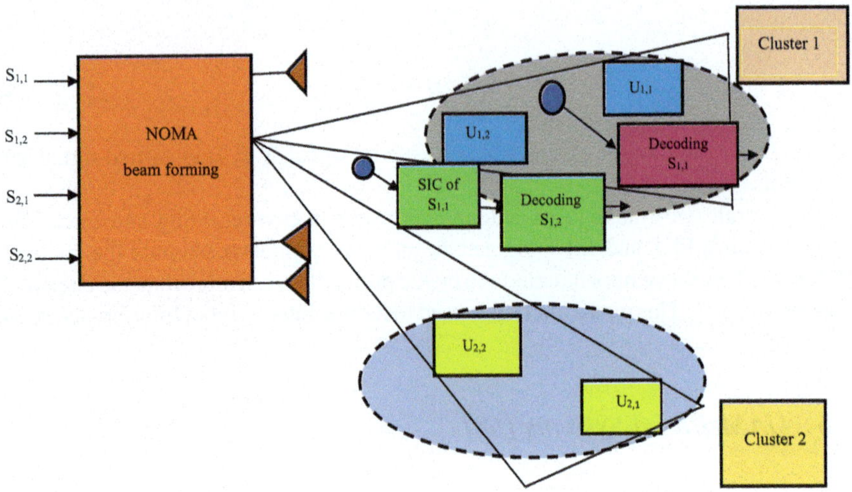

Fig. 3 MIMO-NOMA system

8 Some Experimental Results and Discussions

One of the major factors to look out while designing a 5G and beyond system is the path loss. Also, to make 6G a reality, channel capacity must be enhanced. Considering these factors, path loss, Signal to Noise Ratio (SNR) and channel capacity have been analysed in the Terahertz Band (0.1–10.0 THz).

Path loss depends on transmission distance d and wavelength λ.

$$L_{bf} = 20\log_{10}\left(\frac{4\pi d}{\lambda}\right) \tag{2}$$

Channel capacity (in bps) is dependent on the factors of bandwidth B and Signal to Noise Ratio (SNR) S/N. We know

$$C = B\log_2\left(1 + \frac{S}{N}\right) \tag{3}$$

In Fig. 4, path loss has been analysed in the THz frequency band (0.1 THz-10THz) for transmission distances of 1,10 and 100 m, respectively. It shows that with increase in the transmission distance path loss also increases.

Complementary Cumulative distribution of path loss for the same frequency has been analysed in Fig. 5 for three different transmission distances of 1,10 and 100 m. From Fig. 6 it can be said that Signal to Noise Ratio (SNR) improves for smaller value of the transmission distance.

Within the above-mentioned THz band of 0.1–10.0 THz, channel capacity has been analysed for three different transmission distances of 1,10 and 100 m, which is

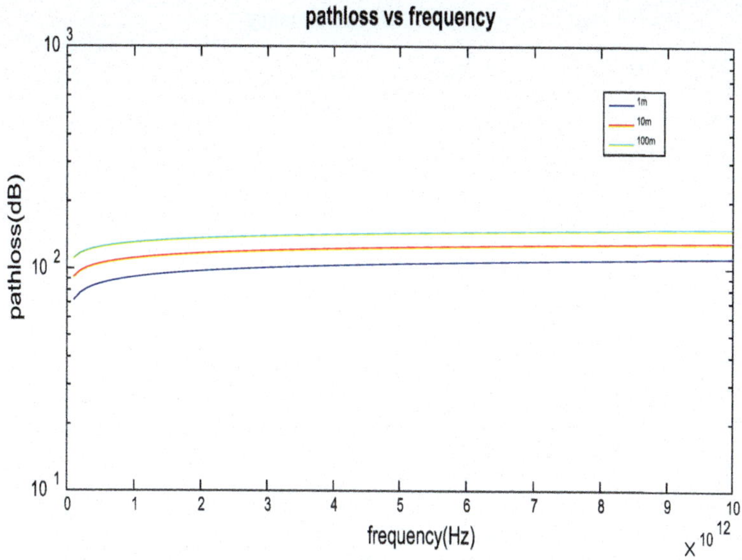

Fig. 4 Path loss vs Frequency at distance 1 m,10 m and 100 m

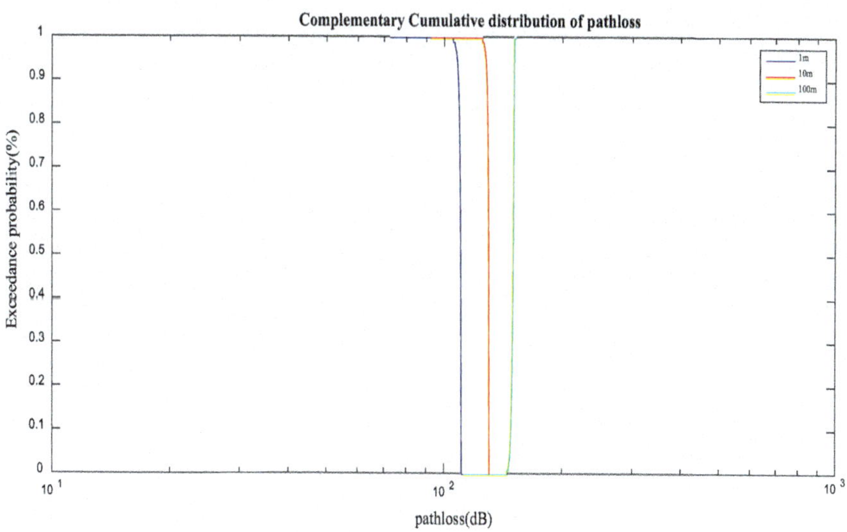

Fig. 5 Complementary Cumulative distribution of Path loss at distance 1, 10 and 100 m

illustrated in Fig. 7. In the same THz band for 100 m transmission distance, maximum channel capacity is 10 Tbps, whereas for 10 and 1 m distances, maximum channel capacity reaches to the value of 40 and 250 Tbps. This clearly indicates that channel capacity increases with increasing frequency.

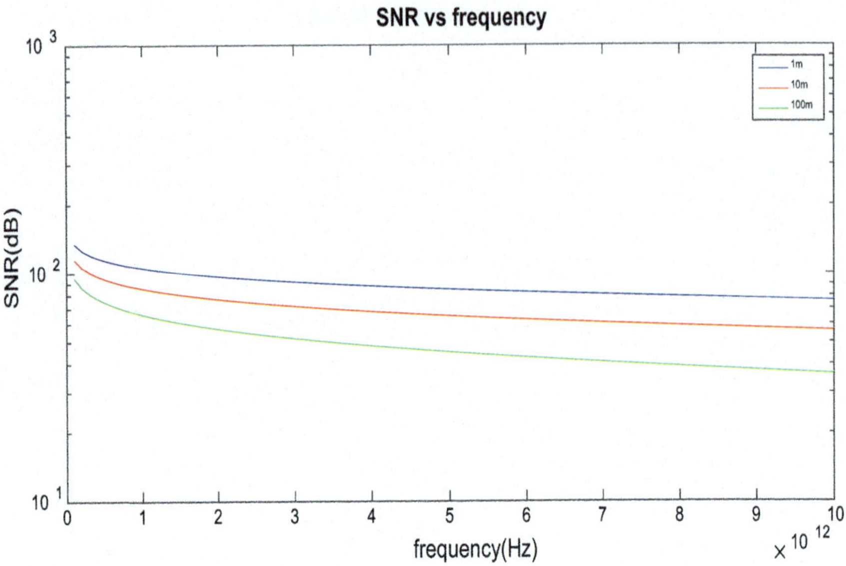

Fig. 6 Signal to Noise Ratio (SNR) vs Frequency at distance 1,10 and 100 m

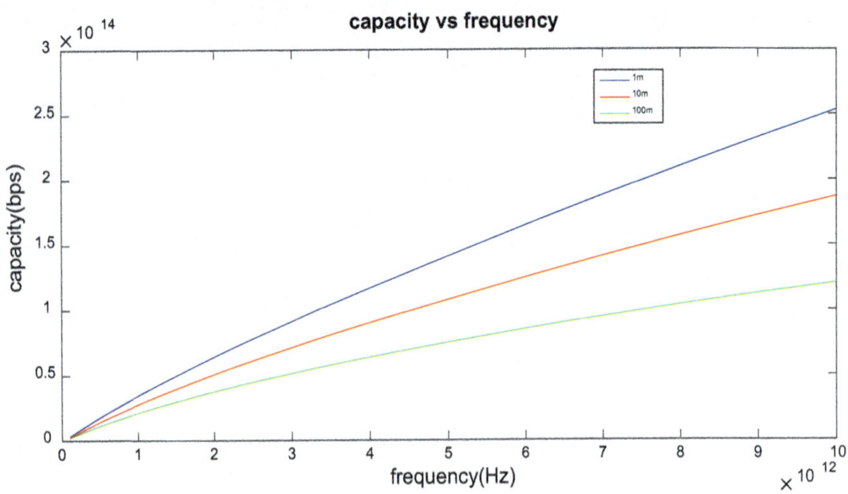

Fig. 7 Capacity vs Frequency at distance 1,10 and 100 m

In Fig. 8, the variation of capacity for the same system is analysed with changing transmission distances for three different frequency values, viz. 0.1, 1, and 10 THz. The graph shows that channel capacity decreases with increasing transmission distance for a particular frequency.

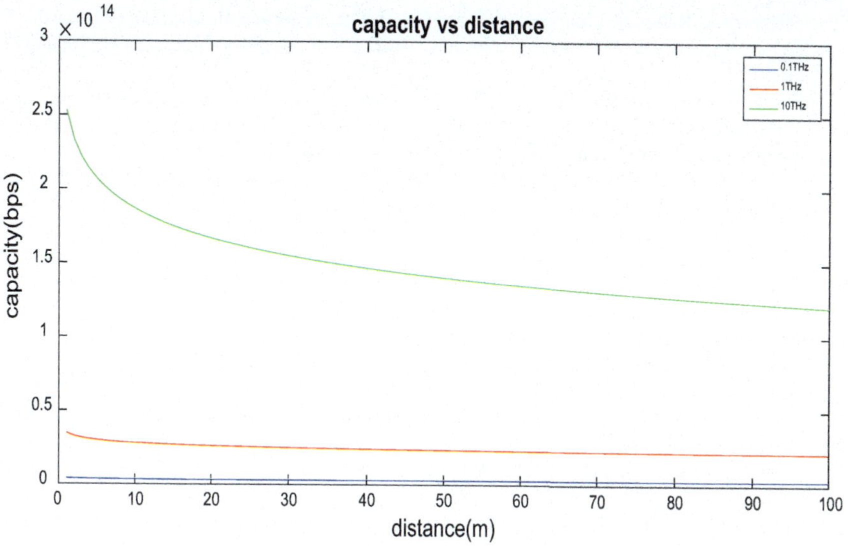

Fig. 8 Capacity vs distance at frequency 0.1, 1, and 10 THz

9 Conclusions

THz is considered as one of the challenging technologies for 6G and beyond communication. Within last 5 years, THz technology has been developed to generate and detect THz radiation. Electric sources, receivers and modulators have been developed for frequencies up to 1 or 2 THz band. Here we discussed various aspects like key features, significance, promising applications and difficulties in THz technology revisiting the recent research publications. To conclude we may say that more research developments are necessary in THz communication and THz devices etc.

References

1. S. Tripathi, N.V. Sabu, A.K. Gupta, H.S. Dhillon, Millimeter-wave and terahertz spectrum for 6G wireless, in *6G Mobile Wireless Networks* (Springer, Cham, 2021), pp. 83–121
2. M. Abdullah, A. Altaf, M.R. Anjum, Z.A. Arain, A.A. Jamali, M. Alibakhshikenari, F. Falcone, E. Limiti, Future smartphone: MIMO antenna system for 5G mobile terminals. IEEE Access **9**, 91593–91603 (2021)
3. S. Mosleh, Y. Ma, J.D. Rezac, J.B. Coder, Dynamic spectrum access with reinforcement learning for unlicensed access in 5G and beyond, in *2020 IEEE 91st Vehicular Technology Conference (VTC2020-Spring)*. (IEEE, May 2020), pp. 1–7
4. H. Ronkainen, J. Edstam, A. Ericsson, C. Ostberg, Integrated access and backhaul. Ericsson Technol. Rev. **6**, 2–11 (2020)
5. J. Peisa, P. Persson, S. Parkvall, E. Dahlman, A. Grovlen, C. Hoymann, D. Gerstenberger, 5G evolution: 3GPP releases 16 & 17 overview. Ericsson Technol Rev. **6**, 2–13 (2020)

6. I. Parvez, A. Rahmati, I. Guvenc, A.I. Sarwat, H. Dai, A survey on low latency towards 5G: RAN, core network and caching solutions. IEEE Commun. Surv. Tutor. **20**(4), 3098–3130 (2018)
7. G. Liu, X. Hou, J. Jin, F. Wang, Q. Wang, Y. Hao, Y. Huang, X. Wang, X. Xiao, A. Deng, 3-D-MIMO with massive antennas paves the way to 5G enhanced mobile broadband: from system design to field trials. IEEE J. Sel. Areas Commun. **35**(6), 1222–1233 (2017)
8. S. Chen, R. Ma, H.H. Chen, H. Zhang, W. Meng, J. Liu, Machine-to-machine communications in ultra-dense networks—a survey. IEEE Commun. Surv. Tutor. **19**(3), 1478–1503 (2017)
9. A. Zaier, R. Bouallegue, Channel estimation study for block-pilot insertion in OFDM systems under slowly time varying conditions (2012). arXiv:1202.1552
10. H. Sarieddeen, M.S. Alouini, T.Y. Al-Naffouri, An overview of signal processing techniques for terahertz communications, in *Proceedings of the IEEE* (2021)
11. R. Piesiewicz, T. Kleine-Ostmann, N. Krumbholz, D. Mittleman, M. Koch, J. Schoebel, T. Kurner, Short-range ultra-broadband terahertz communications: concepts and perspectives. IEEE Antennas Propag. Mag. **49**(6), 24–39 (2007)
12. J. Federici, L. Moeller, Review of terahertz and subterahertz wireless communications. J. Appl. Phys. **107**(11), 6 (2010)
13. X. Huang, Y.J. Guo, J.D. Bunton, A hybrid adaptive antenna array. IEEE Trans. Wireless Commun. **9**(5), 1770–1779 (2010)
14. K.C. Huang, Z. Wang, Terahertz terabit wireless communication. IEEE Microwave Mag. **12**(4), 108–116 (2011)
15. H.J. Song, T. Nagatsuma, Present and future of terahertz communications. IEEE Trans. Terahertz Sci. Technol. **1**(1), 256–263 (2011)
16. T. Kurner, S. Priebe, Towards THz communications-status in research, standardization and regulation. J. Infrared Millim. Terahertz Waves **35**(1), 53–62 (2014)
17. I.F. Akyildiz, J.M. Jornet, C. Han, Terahertz band: next frontier for wireless communications. Phys. Commun. **12**, 16–32 (2014)
18. I.F. Akyildiz, J.M. Jornet, C. Han, TeraNets: ultra-broadband communication networks in the terahertz band. IEEE Wirel. Commun. **21**(4), 130–135 (2014)
19. F. Guo, F.R. Yu, H. Zhang, X. Li, H. Ji, V.C. Leung, Enabling massive IoT toward 6G: a comprehensive survey. IEEE Internet Things J. (2021)
20. N.U. Hasan, W. Ejaz, N. Ejaz, H.S. Kim, A. Anpalagan, M. Jo, Network selection and channel allocation for spectrum sharing in 5G heterogeneous networks. IEEE Access **4**, 980–992 (2016)
21. T. Kleine-Ostmann, T. Nagatsuma, A review on terahertz communications research. J. Infrared Millim. Terahertz Waves **32**(2), 143–171 (2011)
22. T. Nagatsuma, G. Ducournau, C.C. Renaud, Advances in terahertz communications accelerated by photonics. Nat. Photonics **10**(6), 371–379 (2016)
23. I. Mehdi, J.V. Siles, C. Lee, E. Schlecht, THz diode technology: status, prospects, and applications. Proc. IEEE **105**(6), 990–1007 (2017)
24. C. Han, Y. Chen, Propagation modeling for wireless communications in the terahertz band. IEEE Commun. Mag. **56**(6), 96–101 (2018)
25. D. Wu, Q. Wu, Y. Xu, J. Jing, Z. Qin, QoE-based distributed multichannel allocation in 5G heterogeneous cellular networks: a matching-coalitional game solution. IEEE Access **5**, 61–71 (2016)
26. S.L. Nguyen, J. Jarvelainen, A. Karttunen, K. Haneda, J. Putkonen, Comparing radio propagation channels between 28 and 140 GHz bands in a shopping mall, in *Proceedings of 2018 12th European Conference on Antennas and Propagation (EuCAP)* (2018)
27. A.A.A. Boulogeorgos, A. Alexiou, T. Merkle, C. Schubert, R. Elschner, A. Katsiotis, P. Stavrianos, D. Kritharidis, P.K. Chartsias, J. Kokkoniemi, M. Juntti, Terahertz technologies to deliver optical network quality of experience in wireless systems beyond 5G. IEEE Commun. Mag. **56**(6), 144–151 (2018)
28. D. Headland, Y. Monnai, D. Abbott, C. Fumeaux, W. Withayachumnankul, Tutorial: terahertz beamforming, from concepts to realizations. Apl Photonics **3**(5), 051101 (2018)

29. K. Sengupta, T. Nagatsuma, D.M. Mittleman, Terahertz integrated electronic and hybrid electronic–photonic systems. Nat. Electron. **1**(12), 622–635 (2018)
30. I.F. Akyildiz, C. Han, S. Nie, Combating the distance problem in the millimeter wave and terahertz frequency bands. IEEE Commun. Mag. **56**(6), 102–108 (2018)
31. T.S. Rappaport, Y. Xing, O. Kanhere, S. Ju, A. Madanayake, S. Mandal, A. Alkhateeb, G.C. Trichopoulos, Wireless communications and applications above 100 GHz: opportunities and challenges for 6G and beyond. IEEE Access **7**, 78729–78757 (2019)
32. H. Elayan, O. Amin, B. Shihada, R.M. Shubair, M.S. Alouini, Terahertz band: the last piece of RF spectrum puzzle for communication systems. IEEE Open J. Commun. Soc. **1**, 1–32 (2019)
33. Y. Corre, G. Gougeon, J.B. Doré, S. Bicaïs, B. Miscopein, E. Faussurier, M. Saad, J. Palicot, F. Bader, Sub-THz spectrum as enabler for 6G wireless communications up to 1 Tbit/s, in *6G Wireless Summit* (March 2019)
34. W. Haselmayr, A. Springer, G. Fischer, C. Alexiou, H. Boche, P.A. Hoeher, F. Dressler, R. Schober, Integration of molecular communications into future generation wireless networks, in *1st 6G Wireless Summit* (IEEE, Levi, Finland, 2019)
35. O.L. Lopez, H. Alves, R.D. Souza, S. Montejo-Sanchez, E.M.G. Fernandez, M. Latva-Aho, Massive wireless energy transfer: enabling sustainable IoT toward 6G era. IEEE Internet Things J. **8**(11), 8816–8835 (2021)
36. M.J. Piran, D.Y. Suh, Learning-driven wireless communications, towards 6G, in *2019 International Conference on Computing, Electronics & Communications Engineering (iCCECE)* (IEEE, August 2019), pp. 219–224
37. R.A. Stoica, G.T.F. de Abreu, 6G: the wireless communications network for collaborative and AI applications (2019). arXiv:1904.03413
38. D. Kalbande, S. Haji, R. Haji, 6G-Next Gen mobile wireless communication approach, in *2019 3rd International Conference on Electronics, Communication and Aerospace Technology (ICECA)*. (IEEE, June 2019), pp. 1–6
39. S. Ghafoor, N. Boujnah, M.H. Rehmani, A. Davy, MAC protocols for terahertz communication: a comprehensive survey. IEEE Commun. Surv. Tutor. **22**(4), 2236–2282 (2020)
40. K. Tekbiyik, A.R. Ekti, G.K. Kurt, A. Gorcin, H. Yanikomeroglu, A holistic investigation of terahertz propagation and channel modeling toward vertical heterogeneous networks. IEEE Commun. Mag. **58**(11), 14–20 (2020)
41. L. Zhang, X. Pang, S. Jia, S. Wang, X. Yu, Beyond 100 Gb/s optoelectronic terahertz communications: key technologies and directions. IEEE Commun. Mag. **58**(11), 34–40 (2020)
42. M. Polese, J.M. Jornet, T. Melodia, M. Zorzi, Toward end-to-end, full-stack 6G terahertz networks. IEEE Commun. Mag. **58**(11), 48–54 (2020)
43. H. Sarieddeen, N. Saeed, T.Y. Al-Naffouri, M.S. Alouini, Next generation terahertz communications: a rendezvous of sensing, imaging, and localization. IEEE Commun. Mag. **58**(5), 69–75 (2020)
44. F. Lemic, S. Abadal, W. Tavernier, P. Stroobant, D. Colle, E. Alarcon, J. Marquez-Barja, J. Famaey, Survey on terahertz nanocommunication and networking: a top-down perspective. IEEE J. Sel. Areas Commun. **39**(6), 1506–1543 (2021)
45. C. Chaccour, M.N. Soorki, W. Saad, M. Bennis, P. Popovski, M. Debbah, Seven defining features of terahertz (THz) wireless systems: a fellowship of communication and sensing. IEEE Commun. Surv. Tutor. (2022)
46. H. Tataria, M. Shafi, A.F. Molisch, M. Dohler, H. Sjöland, F. Tufvesson, 6G wireless systems: vision, requirements, challenges, insights, and opportunities (2020). arXiv:2008.03213
47. P. Kaushal, 6G wireless communications: vision, research groups, and future. Int. Res. J. Eng. Technol. (IRJET) (2021)
48. S. Elmeadawy, R.M. Shubair, Enabling technologies for 6G future wireless communications: opportunities and challenges (2020). arXiv:2002.06068
49. F. Tariq, M.R. Khandaker, K.K. Wong, M.A. Imran, M. Bennis, M. Debbah, A speculative study on 6G. IEEE Wirel. Commun. **27**(4), 118–125 (2020)
50. G. Gui, M. Liu, F. Tang, N. Kato, F. Adachi, 6G: opening new horizons for integration of comfort, security, and intelligence. IEEE Wirel. Commun. **27**(5), 126–132 (2020)

51. M. Matthaiou, O. Yurduseven, H.Q. Ngo, D. Morales-Jimenez, S.L. Cotton, V.F. Fusco, The road to 6G: ten physical layer challenges for communications engineers. IEEE Commun. Mag. **59**(1), 64–69 (2021)
52. L. Bariah, L. Mohjazi, S. Muhaidat, P.C. Sofotasios, G.K. Kurt, H. Yanikomeroglu, O.A. Dobre, A prospective look: key enabling technologies, applications and open research topics in 6G networks. IEEE Access **8**, 174792–174820 (2020)
53. S. Chen, Y.C. Liang, S. Sun, S. Kang, W. Cheng, M. Peng, Vision, requirements, and technology trend of 6G: how to tackle the challenges of system coverage, capacity, user data-rate and movement speed. IEEE Wirel. Commun. **27**(2), 218–228 (2020)
54. Y. Yuan, Y. Zhao, B. Zong, S. Parolari, Potential key technologies for 6G mobile communications. Sci. China Inf. Sci. **63**(8), 1–19 (2020)
55. Z. Chen, X. Ma, B. Zhang, Y. Zhang, Z. Niu, N. Kuang, W. Chen, L. Li, S. Li, A survey on terahertz communications. China Commun. **16**(2), 1–35 (2019)
56. M. Giordani, M. Polese, M. Mezzavilla, S. Rangan, M. Zorzi, Toward 6G networks: use cases and technologies. IEEE Commun. Mag. **58**(3), 55–61 (2020)
57. A. Mughees, M. Tahir, M.A. Sheikh, A. Ahad, Towards energy efficient 5g networks using machine learning: taxonomy, research challenges, and future research directions. IEEE Access **8**, 187498–187522 (2020)
58. M. Tamosiunaite, S. Tamosiunas, M. Zilinskas, G. Valusis, Atmospheric attenuation of the terahertz wireless networks, in *Broadband Communications Networks-Recent Advances and Lessons from Practice* (2017), pp. 143–156
59. H. Yuan, N. Yang, K. Yang, C. Han, J. An, December. Hybrid beamforming for MIMO-OFDM terahertz wireless systems over frequency selective channels, in *2018 IEEE Global Communications Conference (GLOBECOM)* (IEEE, 2018), pp. 1–6
60. J.F. O'Hara, S. Ekin, W. Choi, I. Song, A perspective on terahertz next-generation wireless communications. Technologies **7**(2), 43 (2019)
61. Y. Chen, Y. Li, C. Han, Z. Yu, G. Wang, Channel measurement and ray-tracing-statistical hybrid modeling for low-terahertz indoor communications. IEEE Trans. Wireless Commun. **20**(12), 8163–8176 (2021)
62. S. Tarboush, H. Sarieddeen, M.S. Alouini, T.Y. Al-Naffouri, Single-versus multi-carrier terahertz-band communications: a comparative study (2021). arXiv:2111.07398
63. H.Q. Ngo, E.G. Larsson, T.L. Marzetta, Energy and spectral efficiency of very large multiuser MIMO systems. IEEE Trans. Commun. **61**(4), 1436–1449 (2013)
64. R.W. Heath Jr., A. Lozano, *Foundations of MIMO Communication* (Cambridge University Press, 2018)
65. T.L. Marzetta, H.Q. Ngo, *Fundamentals of Massive MIMO* (Cambridge University Press, 2016)
66. E. Bjornson, J. Hoydis, L. Sanguinetti, Massive MIMO networks: spectral, energy, and hardware efficiency. Found. Trends Signal Proc. **11**(3–4), 154–655 (2017)
67. A. Omri, R. Bouallegue, New transmission scheme for MIMO-OFDM. Int. J. Next Gener. Netw. **3**(1), 11–19 (2011)
68. Y. Yan, G. Xie, M.P. Lavery, H. Huang, N. Ahmed, C. Bao, Y. Ren, Y. Cao, L. Li, Z. Zhao, A.F. Molisch, High-capacity millimetre-wave communications with orbital angular momentum multiplexing. Nat. Commun. **5**(1), 1–9 (2014)
69. L. Dai, B. Wang, Z. Ding, Z. Wang, S. Chen, L. Hanzo, A survey of non-orthogonal multiple access for 5G. IEEE Commun. Surv. Tutor. **20**(3), 2294–2323 (2018)
70. Y. Huang, C. Zhang, J. Wang, Y. Jing, L. Yang, X. You, Signal processing for MIMO-NOMA: present and future challenges. IEEE Wirel. Commun. **25**(2), 32–38 (2018)

Some Aspects of Novel Materials from Optical to THz Engineering

Swagata Bhattacherjee, Ananya Barman, and Trina Dutta

1 Introduction

Ultraviolet (UV) photodetectors have been effectively implemented in advanced communication systems, air decontamination, ozone sensation, and different types of flame and flaw detection techniques over the past few decades. Continuous development of novel materials synthesis and device fabricating techniques has generated a new version of UV photodetectors, which are reduced in size, consume lower energy, and exhibit higher precision. The journey of Ultraviolet research was on track during the nineteenth century, and it is commonly divided into the following parts.

Near ultraviolet (NUV): It is just next to the visible region, and the wavelength range is 400–300 nm.

Mid ultraviolet (MUV): wavelength is between 200 and 300 nm, and ozone can absorb it.

Far ultraviolet (FUV): wavelength is between 200 and 100 nm, and absorbed by molecular oxygen.

Extreme ultraviolet (EUV): wavelengths belong between 100 and 10 nm. In this range, atomic, and molecular gases become dominant absorbers.

Apart from the above names, the International Commission on Illumination (CIE) has classified the UV spectrum into different bands [1] as shown in Table 1.

UVB and UVC are the strongest and destructed for human beings, particularly at 270 nm. In this wavelength the UV contact limitation of a living person become very low, around 3 mJ/cm^2 in one second [1] (Fig. 1).

S. Bhattacherjee (✉)
Department of Physics, JIS College of Engineering, Kalyani, Nadia, India
e-mail: swagata.bhattacharjee@jiscollege.ac.in

A. Barman · T. Dutta
Department of Chemistry, JIS College of Engineering, Kalyani, Nadia, India

A. Saha et al. (eds.), *Optical to Terahertz Engineering*, Progress in Optical Science and Photonics 23, https://doi.org/10.1007/978-981-99-0228-6_5

Table 1 Different bands of the Solar radiation spectrum

Name of the band	Wavelength range
Vacuum ultraviolet (VUV)	Below 180 nm
Deep ultraviolet (DUV)	220–280 nm
Ultraviolet (UVA)	320–400 nm
Ultraviolet (UVB)	280–320 nm
Ultraviolet (UVC)	100–280 nm

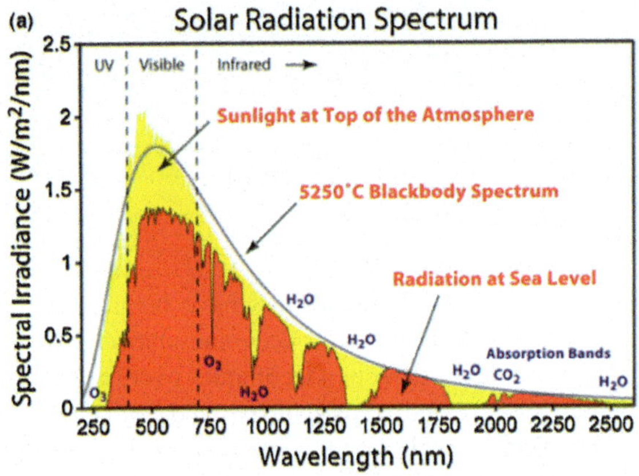

Fig. 1 Solar radiation spectrum [1]

Traditional UV photodetectors are stable, highly sensitive, and fast having high signal-to-noise (S/N) ratio. But the performances of traditional UV photodetectors need to be revised to satisfy the upward necessities and opportunities. Thus the technology is shifted towards novel UV photodetectors implementing some special features like smartness, flexibility, and intelligence. Although Si is the base material for the fabrications of the present semiconductor UV detectors, in many cases the applications for UV detection involve intimidating environments like situ combustion monitoring or satellite-based missile plume uncovering, where the harshness of vigorous material plays a significant role.

Whereas in other cases like air quality observing or gas detecting, the sensitivity of the material is more crucial. In aerospace, automotive, and petroleum industries requirements for fringe technologies are increasing in which high temperature tolerant and unfriendly surroundings are required. Thus in the regime of optical devices, research activities are also going toward novel materials in order to fulfill different kinds of requirements. In Sect. 2 some conventional materials are discussed followed by some novel materials in Sects. 3 and 4, which are extensively utilized for photo-sensing and detection from visible to UV regions.

2 Wide Bandgap Materials

These are usually wide band gap semiconductors, where the photon is allowed to interact with the semiconductor layer to achieve photoelectron in the vacuum environment. A voltage is incorporated between the cathode and an anode causing a photocurrent to flow, depending on incident radiation intensity. For this kind of detector, for which the working principle is based on the photoelectric effects, the selection of wavelength range is mainly governed by the work function of the substrate. For semiconductor-based detectors, semiconductor material producing electron–hole pairs which are separated by the application of an electric field absorbs photons. For semiconductor photodetectors, the working principle depends on the internal photoelectric effect where the incident radiation is sufficiently large to transit the electrons from the valence to the conduction band.

The three fundamental modes of working of ultraviolet semiconductors are as follows:

- photoconductive detectors,
- p-n junctions photodiode, and
- metal–semiconductor-metal (MSM) detectors.

Photodetectors in the UV region are fabricated by different wide-band semiconductor compounds.

2.1 *Silicon Carbide (SiC)*

SiC is considered for fabricating high-temperature, high-power electronic devices including photodetectors due to its wide band-gap energy [2]. SiC-based photodetectors exhibit large gains, high signal-to-noise ratio, and solar-blind operation, for which it is widely used in UV spectrum monitoring without using solar rejection filters. Moreover, SiC offers exceptional long stability under high-intensity UV radiation and high operating temperature. Using rapid thermal chemical vapor deposition (RTCVD) techniques a hetero-epitaxial SiC/Si MSM photodetector is developed for high-temperature deep-UV detection as shown in Fig. 2a [2].

The performances of the SiC-MSM photodetectors were examined by the measurement of photo and dark currents and their ratio under various operating temperatures. The obtained dark current is less than 0.12 μA at the applied voltage of 5 V and 0.8 μA at 10 V at room temperature [2]. An increase in temperature up to 80 °C causes an increase in the dark current. The possible reason for obtaining a higher dark current is the large doping of SiC which is used to get high responsivity. On exposure to 250 nm illumination with an intensity of 2 mW/cm^2, the photocurrent was obtained as 55 μA at the bias of 10 V at room temperature as depicted in Fig. 3a.

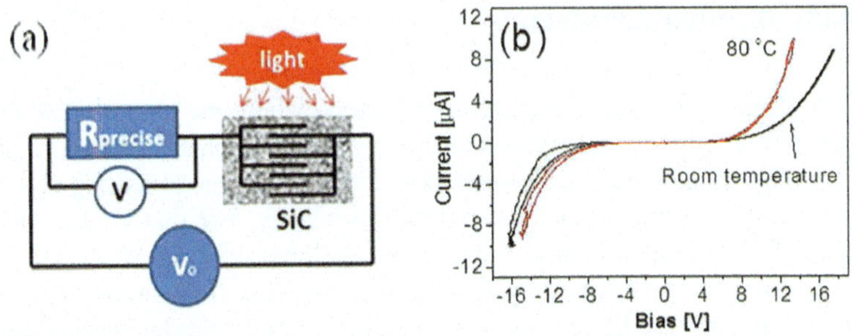

Fig. 2 **a** Schematic diagram of the experimental setup for characterization of responsivity of SiC-based photodetector electric properties and **b** measurement of dark current in different temperatures with bias voltage [2]

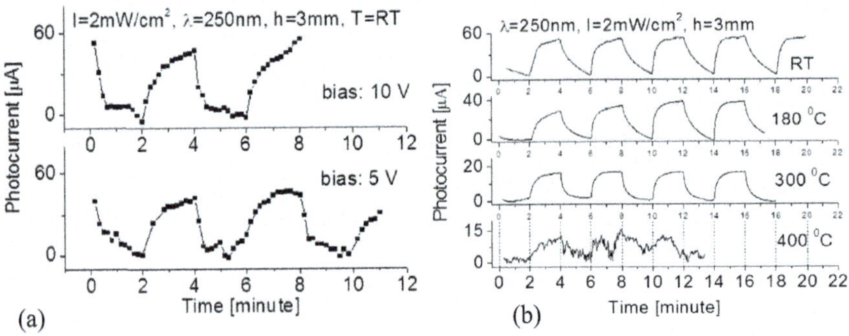

Fig. 3 **a** Bias dependence of photoresponse at room temperature **b** Photoresponse in different temperature for λ = 250 nm light with intensity of I = 2 mW/cm^2 at room temperature [2]

Figure 3b represents the photocurrent with the same intensity and wavelength of incident radiation at different temperatures. The figure shows that the output baseline of the SiC-based photodetector is very stable and well-defined up to 400 °C.

2.2 *Gallium Nitride (GaN)*

The III–V nitrides are very attractive members of the UV range photo device. Advancements in fabrication technology make these materials a natural choice in the infrared, red, and green wavelength domains. But the development of suitable technology with this material in the UV spectral region is limited. The III–V nitrides are excellent materials having a band gap of ~ 4.75 eV which is compatible with the 240–280 nm technology band, where ozone absorption takes place which makes the earth's atmosphere less transparent [3]. These materials are better candidates than

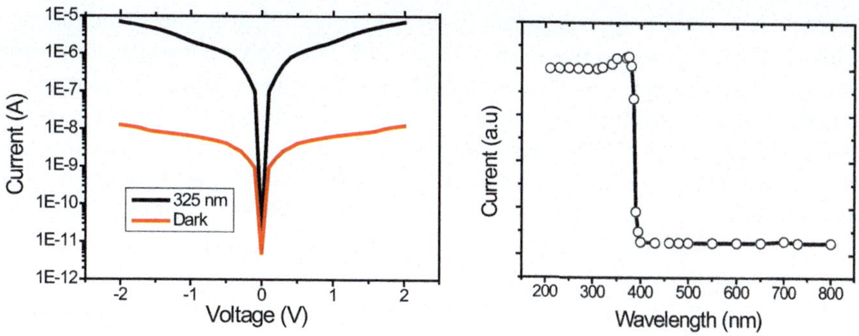

Fig. 4 **a** I-V characteristics of a nanowire (NW) photodetector under dark and illumination and **b** response in different wavelengths [3]

silicon regarding the hardness characteristics of radiation. Among different III–V nitrides, GaN shows the best performance, especially in the UVA range.

The high absorption coefficient and direct wide band gap made GaN one of the most capable materials for detecting the UV spectrum. Figure 4a depicts the I-V characteristics of an Ag/single nonpolar GaN NW detector [3]. The black curve represents the dark current while the red curve shows the currents for 325 nm UV light illumination with a power density of 1 mW/cm^2. The curves indicate that the current across the NW increases sharply under light illumination. Figure 4b shows a photon-response spectrum of the device under the wavelength range of 200–800 nm at a bias of 1.0 V. It is found that the response is almost constant except for the frequency range 370–385 nm which is related to the band gap of GaN. In this domain, a sharp change of response is observed, which ensures the perfect wavelength selectively of the material. GaN NW is not only used for UV photodetection but also for stretchable photosensors. Stretchable photosensors are getting attention for various applications like biological sensing, image sensing, health-monitoring systems, and wireless communication. These photosensors show very high photocurrent and photoresponsivity.

Although a series of experiments focusing on nanowire UV detectors have been performed with GaN [4], still the applications of GaN in UV photodetection are limited due to the lack of high-quality materials. Heavily n-type doped GaN suffers from native defects caused by nitrogen vacancies whereas, p-type doping is challenging. In course of technological development, these limitations are also mitigated gradually.

2.3 *Zn-Based Materials*

Zinc oxide (ZnO) is a well-known direct band gap material to fabricate ultraviolet (UV) photodetector owing to its band gap, which makes it suitable for high absorption as well as high rejection in the UV range [5]. It is cheap and biocompatible. Also, ZnO-based photosensors offer comparatively high resistance to irradiation in comparison with diamond, SiC, and GaN. Depending on the mode of operation, ZnO-based photodetectors are classified as photoconductive type, metal–semiconductor-metal (MSM) type, and p–n junction type [6, 7]. P–N junction-type photodetectors have several advantages over the other two. It offers low dark current, high response speed, and broad responsivity in the U-V range. ZnO NW photodetectors and optical switches are also getting much attention and being the subject of extensive investigations.

Due to the large surface-to-volume ratio and the presence of deep-level surface trap states, these nanostructures contribute to high photosensitivity [8]. The downsizing of the active area also shortens the carrier transit time resulting in substantial photoconductive gain [9].

Figure 5a shows the variation of photocurrent with the intensity of light whereas Fig. 5b represents the photoconductive gain at an applied bias of 5 V. As shown in the figure, the photocurrent increases linearly with light intensity, when the intensity is small. But at higher light intensities, a lack of linearity is observed and the sublinear dependence of the photocurrent on light intensity indicates the reduction of available hole traps near the surface at higher illumination leading to the saturation of the photoresponse. Figure 5b shows a good photoconductive gain of the device which is more than 10^8 [5].

ZnS is also one of the direct wide bandgap semiconductors that provides a novel perspective alternative for UV detectors that is compatible with the UV-A band. It has cubic zinc blend (ZB) as well as hexagonal wurtzite (WZ) structures with a wide band gap of 3.72 and 3.77 eV respectively [10]. It is chemically stable and has a higher potential as a UV detector in this specific wavelength regime in comparison to other wide-band gap semiconductors. But ZnS nanostructures usually execute weak photocurrent and are only sensitive to UV light with a wavelength shorter than 335 nm. On the other hand, ZnO-based UV photosensors have good contact with electrodes and show a large photocurrent. But it suffers from poor photocurrent stability caused by oxygen vacancies and zinc interstitial defects. Thus by assembling unique

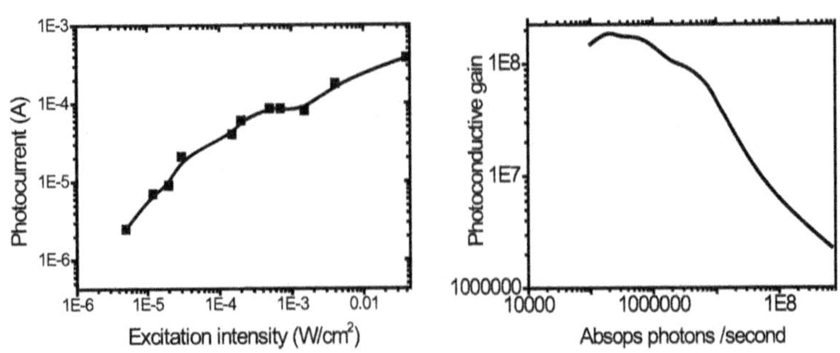

Fig. 5 **a** Photocurrent and **b** photoconductive gain of a ZnO-NW photodetector [5]

features of individual constituents, hybrid ZnS-ZnO can exhibit outstanding operational characteristics, tunable spectral selectivity, fast response speed, and excellent environmental stability in the 'visible-light-blind' or 'solar-blind' ultraviolet (UV) domain with a high degree of precision and great accuracy [11, 12]. ZnO-GaN combination also works well in U-V photo-detection since GaN and ZnO have the same crystalline structure. A high-quality heterojunction is obtained due to a very small lattice constant mismatch of two [13]. In Fig. 6a, a ZnO/GaN-based photodetector is illustrated.

Here the top and the bottom layers are made with an n-ZnO and a p-GaN respectively forming a heterojunction diode, working as a photodetector. As the GaN (3.4 eV) has a little bit larger band gap compare to ZnO (3.37 eV), the GaN layer not only serves as a p-region but also works as a "filter" (transparent beyond 360 nm wavelength) in the back illumination. Figure 6b presents the absorption and transmission spectra of the n-ZnO and p-GaN layers. From the figure, it is found that the GaN layer has a sharp transmission edge at 360 nm whereas the absorption spectrum of the ZnO layer shows a peak value at 366 nm. Also, an overlapped region of two curves is observed between the wavelength 360–390 nm which indicates a high spectrum selectivity of this combination. Thus ZnO/GaN heterojunction photodetectors reflect a sharp precise response span within 360–390 nm.

Zinc selenide is another kind of II–VI compound semiconductor that has drawn considerable attention as a material for prospective optoelectronic devices. ZnSe has a direct bandgap of 2.70 eV which is compatible with 460 nm at room temperature and widely applied in II–VI light-emitting diodes as well as semiconductor lasers in the blue domain [13].

A nanobelt is a 1D nanostructure-based channel material which has a definite chemical composition, crystallographic orientation, and encircling surfaces that work as performance boosters [14]. Nanobelt functions as nano-scale multiband photo sensors when exposed to light [15]. Individual ZnSe-nanobelt photodetectors are

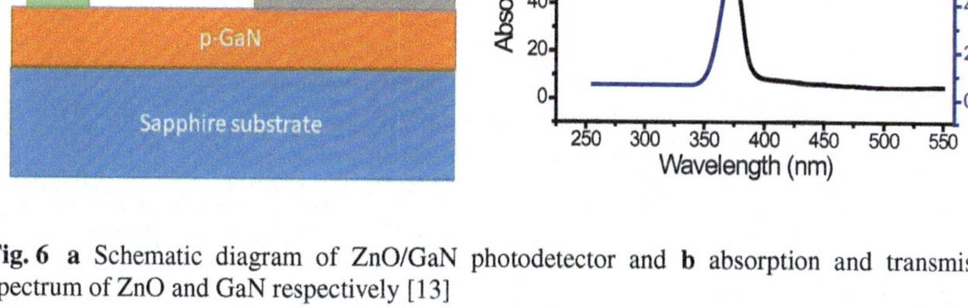

Fig. 6 **a** Schematic diagram of ZnO/GaN photodetector and **b** absorption and transmission spectrum of ZnO and GaN respectively [13]

well accepted as blue/UV-light-sensitive detectors and ultrafast switching devices specifically can detect very low dark current which is beyond the detection range of the measuring instrument. It also exhibits a high photocurrent immediate decay ratio (>99%), and a faster response time (<460 nm) [13].

Figure 7a depicts the schematic diagram of a ZnSe nanobelt photodetector, and Fig. 7b presents its I-V characteristics under the illumination of 400 nm. The graph reveals that the dark current of the photodetector is very low (~10^{-14} amp) [13] even beyond the range of the measuring instrument whereas the photocurrent of 7 pA is noted when a voltage of 30 V is applied for 400 nm light. By illumination with radiation of 500 nm, the photocurrent is found to be increased by two orders of higher magnitude when the device is fed with a voltage just above the threshold excitation energy [13]. Another notable feature of the nanobelt photodetector is that it offers outstanding wavelength selectivity. The interesting properties of the ZnSe photodetector are summarized in the Table 2.

The wide bandgap $Mg_xZn_{1-x}O$ alloy has gained considerable attention owing to its wide modulating bandgap, varying from 3.37 to 7.8 eV. Its growing demand makes it a capable member of the deep ultraviolet (DUV) optoelectronic family. A range of MgZnO-based photogenerators and detectors are reported in which some interesting progress is noticed especially in the field of UV detection [16]. The remarkable

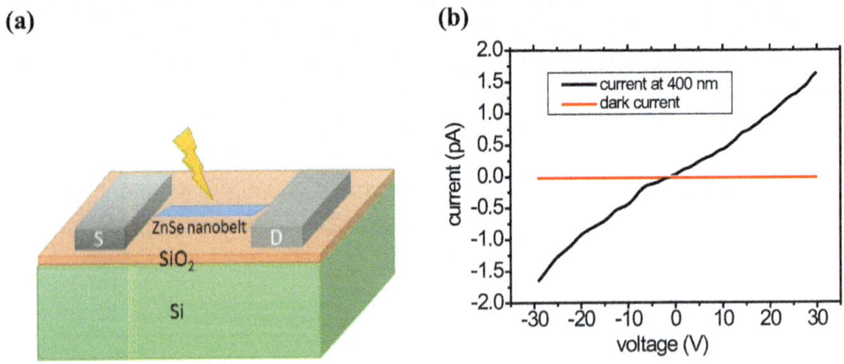

Fig. 7 **a** Schematic diagram of ZnSe nanobelt photodetector **b** I-V features under the radiation of 400 nm [13]

Table 2 Different properties of ZnSe nanobelt photosensor [13]

Parameters	Value
The spectral responsivity	$0.12AW^{-1}$
External quantum efficiency	37.2%
Dark current	<0.01 pF
The ratio of photocurrent to immediate decay	99.4
Response time	<0.3 s

feature of mixed-phased (MP) MgZnO is that it consists with dual-band gaps. The first one is equivalent to the solar blind (220–280 nm) and the second one belongs to the visible blind (300–360 nm). The result is that MP-MgZnO can apply to detect both the solar blind and visible blind, one time in a single detector. However, MgZnO often suffers from phase segregation during the growth process or post-annealing, and it is still a limiting factor for practical applications [16].

3 Novel Material

3.1 Graphene Channel Photodetectors

Apart from wide band gap materials, a wide range of new materials have been familiarized to expand the wavelength range availability and performance enhancement of photonic devices. 1D nanostructures already performed successfully and offers advanced photosensitivity, and have become the natural choice of research over the years. Like 1-D nanomaterials, 2D nanomaterials also have interesting unique features and are considered the prospect of the next-generation photodetectors. Quantum confinements and the absence of interlayer interactions of 2-D nanomaterials make the device smaller, flexible, and efficient. These 2-D nanomaterials are also gaining attention in photodetection by means of prominent surface effects caused by large surface-to-volume ratios.

Graphene is a 2-D material system consisting of one atomic layer of carbon atoms arranged in a regular hexagon. It is a zero-band gap material with a linear energy dispersion relation. It has almost negligible effective mass, having a linear energy–momentum relation. As a reason for this, it possesses high Fermi velocity (~1/300 of light) and large mobility [17]. As the response time is controlled by carrier mobility, graphene-based photodetectors work very fast. Graphene also shows outstanding optical properties. Despite being a single atom thick, it has a very strong inter-valley transition and it can absorb photons from the visible to the THz range [18] for which, graphene-based PDs could be applied over a much wider wavelength range. These optical transitions can be significantly improved in presence of an electric field via a gate, which offers a valuable dimension for optically probing graphene band structure. In the case of multi-layer graphene, absorption takes place cumulatively and the absorption range can be modulated by altering the external gate bias [19].

The strong and optical transitions between different layers of graphene, and its changing behavior in presence of gate bias hold good potential for infrared optics and optoelectronics applications. It is observed that large-area mono or multi-layers of graphene FETs can serve as ultrafast photodetectors. By light illumination, the photogenerated carriers recombine within tens of pico-seconds [20] which are separated in opposite directions in presence of the internal or external field. In Fig. 8a a graphene photodetector is demonstrated where internal fields are used to generate an ultrafast photocurrent response [21]. Here no direct biasing is required in between the source and drain as the carrier transport velocity is very high even under a moderate electric field. Figure 8b represents typical DC characteristics with respect to the source-drain voltage in darkness as well as in presence of a fixed gate bias. It is observed that the location of optical illumination has a strong impact on the magnitude of the photocurrent for constant gate bias.

The maximum photocurrent is obtained for the illumination near the metal-graphene interfaces where the strongest electric field is formed. However, at a sufficiently large positive gate bias, the maximum electric-field transfers towards the center of the graphene channel. The device can produce 6 to 16% internal quantum efficiency within the photo-detection region [21].

In Fig. 9, a novel hybrid graphene–quantum dot phototransistor is depicted which possesses high photodetection gain and quantum efficiency [22]. It consists of a monolayer or bilayer graphene sheet that is sensitized with colloidal PbS quantum dots. Here Graphene is used for carrier transportation in the channel, whereas the function of quantum dots is photon absorbing. The diagram presents the corresponding photo response over a large area upon illumination. Light absorption in PbS quantum dots causes photo generation. The electron–hole pairs are separated at the graphene/quantum dot interface by the induced internal field which results in a band bending near the interface due to the work-function mismatching of graphene and the quantum dots. Trapping of electrons in the quantum dots generates the induced positive carriers in the graphene sheet through capacitive coupling and as long as the electrons remain trapped in the quantum dot, positive charges in the graphene sheet are regenerated which ensures the device gain.

Although ultrafast photodetection is possible for 2D nanomaterials using pristine graphene photodetectors, the responsivity is quite poor [23] and the reason is the zero bandgap and short photo carrier lifetime of graphene, and its practical applications in optoelectronics are still inadequate. By incorporating a band gap, Graphene could be made luminescent in the following two possible ways. The first approach is by

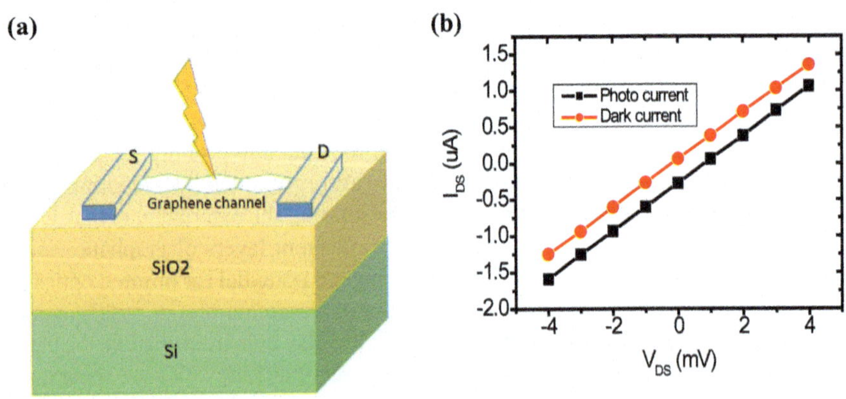

Fig. 8 **a** Schematic diagram of graphene photodetector **b** I-V characteristics under dark and illumination [21]

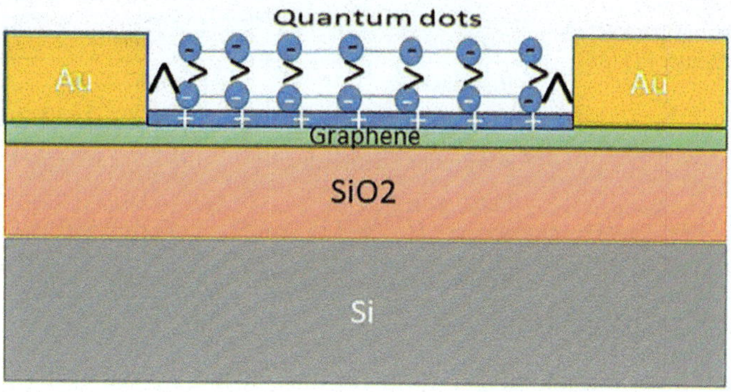

Fig. 9 Schematic diagram of a hybrid graphene quantum dot (PbS) photodetector [22]

cutting it into ribbons and quantum dots [23], and the latter one is by reducing the connectivity of the π electrons network by chemical or physical treatments [24]. Under mild oxygen plasma treatment, discrete graphene flakes become luminescent [24]. Hybrid structures of graphene are also being formed by etching just the top layer, keeping the underlying layers unharmed [24]. These combined layers could be applied in LEDs in the infrared, visible, and blue spectral range. By using plasmonic nanostructures, nanoparticles, or microcavity structures, the optical field absorption can be improved significantly and an enhancement of the responsivity of graphene-based PDs is obtained [25].

3.2 *Layered MoS_2 Photo-Detector for Harsh Environment*

Tsai et al., designed a device by using a 2D layered MoS_2 such as metal–semiconductor–metal photodetector (MSM PDs), useful for the harsh environment. In their research work, they observed that the device exhibited high sensitivity, high thermal stability, and ultra-high quick responsively from visible to UV. In research work, the CVD (chemical vapor deposition) method generally was applied to synthesize layered MoS_2 [26, 27]. Tsai et al., in their research, synthesized layered MoS_2 by using the thermolysis process [28]. 0.25 g $(NH_4)_2MoS_2$ was added to the DMF making a 1.25wt% solution and sonicated for 20 min. After that again it was dipped into the $(NH_4)_2MoS_2$ followed by being heated to 120 °C for 30 min. After that, the mixture was placed into a quartz tube in presence of H_2 where it was heated from 500 to 1000 °C, initiating the formation of the layer of MoS_2 on the sapphire surface.

Finally, layered MoS_2 was kept on the SiO_2/ Si for the fabrication of the photodetector device. The AFM study showed the formation of three layers of MoS_2 of thickness around 0.65 nm for each layer. From HRTEM analysis it was observed that the synthesized MoS_2 layers were highly crystalline with the hexagonal symmetry arrangement of Mo atoms. Ramon spectra and optical microscopic data confirmed

Table 3 Various parameters of MoS2 Schottky PD [28]

Parameters	*Value*
Responsivity	0.57 A/W
Detectivity	10^{10}cmHz$^{1/2}$/W
Absorption coefficient	7.5×10^5 cm^{-1}
Photo gain	13.3
Response times	70 μs
Recovery times	110 μs

the uniform layer formation. The response time of the layered MoS_2 device was calculated as 70–100 μs which was shorter than another device [29]. From the optical absorption spectra, it was observed that layered MoS_2 can absorb around 10% of the incident light. The enormous development of photoluminescence (PL) has been noticed for the monolayer MoS2 both experimentally and theoretically in the current literature [30].

The bandgap expands from 1.2 eV (bulk) to 1.9 eV (monolayer) as a result of quantum confinement and a transition from "indirect bandgap" to "direct bandgap" [30, 31] was taken place. This strong photoluminescence of ultrathin MoS2 has a huge prospect in optoelectronics devices like PDs, biomedical markers, and sensors [31]. It is observed that a few-layer MoS2 Schottky PDs is suitable for photo-detection from visible range to UV regions in harsh environments in which temperatures can increase up to 200 °C. It was also found out that if the layered MoS_2 device is designed in MSM manner, where the Au layers as Schottky contact on the surface of MoS_2, the photosensitivity increases lowering the dark current.

The following table presents different parameters of MoS2 Schottky PDs at 200 °C temperature [28] (Table 3).

3.3 *Monolayer Nano Nets*

In another research work, polymer colloids were also used for the preparation of SnO_2 nano nets, which were used as a photodetector. The advantage of this method is the holes of nano nets which are adjusted by the size of colloidal polymer spheres. The semiconducting oxides like ZnO, TiO_2, or CeO_2 may be used to make the nano nets. But Chen et al. [32] for the first time, used SnO_2 nano nets to design the photodetector device which showed excellent sensitivity and reproducibility. In their research work, Chen et al. [32] fabricated the monolayer colloidal polymer nanofilm by mixing poly (styrene-co-acrylic acid) (PSA) in water and then hexane was added to form a hexane/water interface. Next, the addition of a little amount of ethanol with the help of a syringe, at once the spheres of colloidal polymer were transported to the interface Fig. 10.

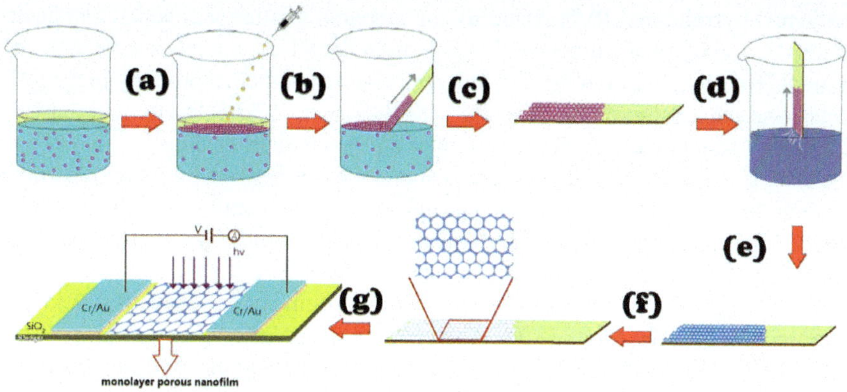

Fig. 10 Fabrication of monolayer SnO_2 nano net-based device as photodetector [32]

An ordered and close-packed monolayer nanofilm of polymer colloidal spheres was formed due to the surface pressure compression. Finally, a solution of SnO_2 was dispersed into the polymer nanofilms to fill the voids and calcined at 600 °C.

SEM images showed that the diameter of formed colloidal nanospheres was 308 nm and the film thickness is equal to the diameter of the spheres. From the SEM data, it was also observed that the film showed a good long-range 2D order. After the insertion of SnO_2 in the nanofilm SEM analysis showed that the presence of 251 nm hole size was smaller than the polymer in the nanofilms having 6 and 94 nm thick walls [32]. The concentration of precursor controlled the formation of ordered and discontinuous nanofilm XRD data analysis confirmed also the formation of high crystalline SnO_2 nanoparticles.

To design the SnO_2 nanofilm photodetector the nanofilm was transferred to an electric gun followed by the deposition of Cr/Au electrodes. The SnO_2 nanofilm was connected with the electrodes and placed around 30 μm apart from each other. While the device was illuminated by light with the wavelength 320 nm at a low voltage of 1.0 V, increased by 3.5 times higher than the dark condition due to the excitation of electron–hole pairs [32]. Thus SnO_2 incorporated nano film-based photodetector showed excellent sensitivity, stability, and reproducibility.

3.4 Bi_2S_3 Nano-Belts

The novel material of low dimensional Bi_2S_3 can also be used as a photodetector which is stable and responds quickly. Li et al. in their research work [33], used Bi_2S_3 nano belts as the light sensors for the devices named surface acoustic wave (SAW) device. The sensor was fabricated on ST-cut quartz by photolithographic technique. For the synthesis of Bi_2S_3 nano belts, triphenyl bismuth, and dibenzyl disulfide was mixed with 4 ml oleylamine (OLA). After that the ethanolic solution of 0.4 g polyvinyl pyrrolidone (PVP, Aladdin), the prepared mixture was added and heated at 180 °C for 8 h in an autoclave. Li et al., used the ST-cut quartz as a piezoelectric material for its unique stability and low-temperature coefficient property in the SAW delay-line device. On this device, the sensitive layer of Bi_2S_3 nano materials was formed by spin coater as a SAW resonator and photodetector, Figs. 11 and 12a. In the analysis of SEM data, it was observed that the synthesized Bi_2S_3 nano particles were uniform, porous, and belt-shaped and the average length was 5–20 μm. XRD showed that the peaks only correspond to the orthorhombic Bi_2S_3 nano particles, indicating the formation of pure Bi_2S_3 nano particles.

It was evident from HR-TEM images that the spacing of the lattice was 0.353 nm corresponding to the plane (310). There was a broad peak observed in UV–VIS spectra between 300 and 700 nm indicating clearly the synthesized nano particles can be used as photodetectors [33]. In their research work, it was observed the transmission signal of SAW before and after the fabrication with nano particles

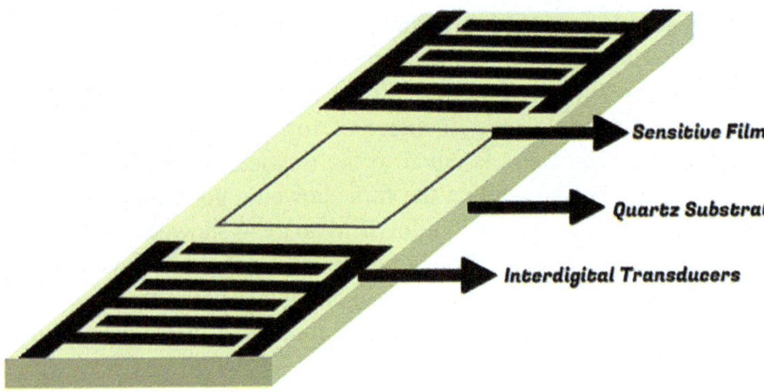

Fig. 11 Fabrication of light sensor of Bi_2S_3 nano belts on quartz [33]

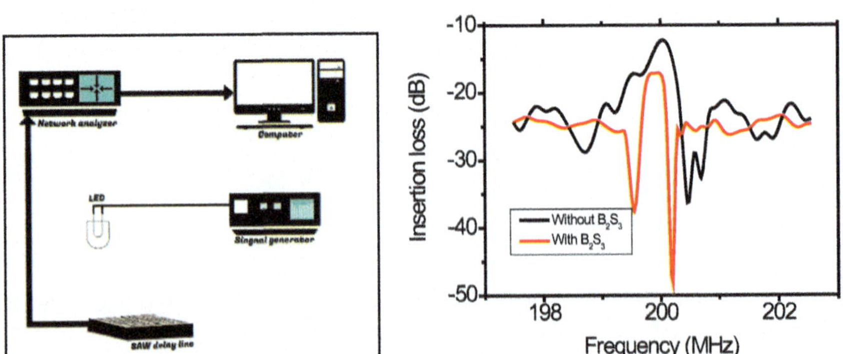

Fig. 12 **a** Experimental setup [33] **b** Transmission signals of SAW device with or without Bi_2S_3 [33]

showed that the central frequency of 200.02 MHz which was set earlier as fabricated, decreased and increased the insertion loss due to the diffraction of the wave by Bi_2S_3, as shown in Fig. 12b. While the light was switched on, the Bi_2S_3 nano belts caused a positive shift of the frequency, and the recovery of the working frequency of took place after the SAW device was turned off. It was observed that for these 3 repeated cycles the same dynamic response curve was found for the same light intensity confirming the fact that the nano bets were reproducible. If the light intensity will be increased, the frequency shift was also increased. If the intensity of the light increased to 170 μWcm^{-2}, around a 7 kHz shift of frequency was taking place. The sensitivity (S) can be determined by the below equation,

$$S = (\Delta f / f) / P \quad (1)$$

where Δf = shift of frequency, f = resonant frequency of device P = power density of incident light. It has been found after calculation, that these nano belts showed good sensitivity in presence of the illumination of visible light. Li et al. proposed that oxygen molecules were absorbed in the porous site of Bi_2S_3 nano belts, which was responsible for the photo response of nano particles [33]. In their research work, they suggested that the photo response was based on the adsorption of O_2 and desorption of O_2 on the nano belt's surface. In dark conditions, the O_2 molecule first captured the free electrons present in the nano belts forming an anion O_2^-.

$$O_2 + e \rightarrow O_2^-$$

After that when the nano belts were irradiated by visible light, some pairs of electron holes were generated where the O_2^- was discharged. This discharge of O_2^- led to the propagation of acoustic waves, increasing the central frequency. Thus increase in light intensity, causes more oxygen release and a large frequency shift was observed.

$$O_2^- + h^+ \rightarrow O_2$$

Thus the novel material, Bi_2S_3 nano belts can be used as an excellent photodetector.

3.5 *$NiCo_2O_4$ Nano-Film*

Hu et al., for the first time, introduced the $NiCo_2O_4$ Nano film as a promising photodetector due to its suitable band gap around 2.1 eV [34], quick response, and high stability. In this mixed valence oxide, Ni is in an octahedral structure and Co shares both octahedral and tetrahedral sites [35]. In their research work, they used the self-assembly of $NiCo_2O_4$ Nano film in oil or water interface as a photodetector. Co_2Ni_1-LDH (layered double hydroxide) was used to synthesize the $NiCo_2O_4$ platelets by

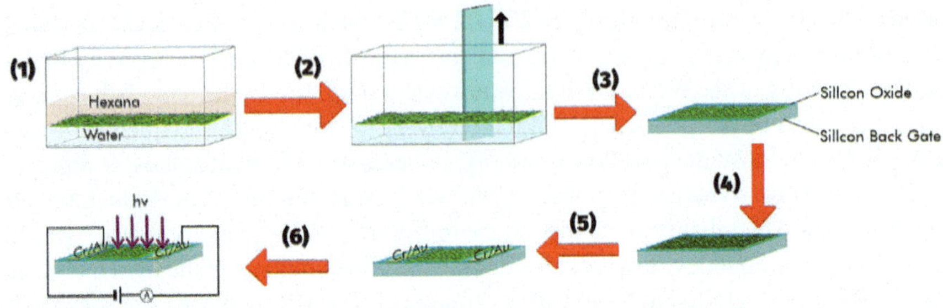

Fig. 13 Schematic diagram for the formation of nano film of $NiCo_2O_4$ [38]

the thermal transformation process. After that, the platelets were dispersed into water and a monolayer nano film of $NiCo_2O_4$ was formed at the interface of hexane/water through the self-assembling method, Fig. 13. There were some interspaces in the monolayer film which can interfere with the propagation of carriers and decrease the efficiency of the nano film. Hence Linfeng Hu et al., developed a two-layer of nano film by repeating the deposition method to remove the interspaces.

TEM images showed that the $NiCo_2O_4$ platelets were hexagonal with sharp edges. From AFM data analysis showed that the average film thickness was around 70 nm. The selected area electron diffraction (SAED) showed sharp diffraction to the plane [111], indicating that the formed $NiCo_2O_4$ platelets were of a single crystal with no domain boundaries. The top surface corresponded to that plane [111]. All this data indicated also that the Co and Ni were distributed uniformly in the crystal lattice. The XRD of nano film confirmed that there was no impurity in the crystal of $NiCo_2O_4$. From SEM analysis it was evident that the nano film was highly oriented with a large coverage area. The device of $NiCo_2O_4$ platelets nano film was designed by the electron beam deposition method.

At first, the self-assembly at the interface was kept onto the SiO2/Si wafer and then Cu/Au electrodes were deposited to design the device of top contact. The electrodes were approximately 50 μm apart from each other. While the device was illuminated with light having energy higher than 2.10 eV, of 589 nm or white light, the current was increased. The photocurrent was increased by 2.3 and 5.0 times larger than the dark current in presence of the light of 550 nm wavelength due to desorption of O_2 which led to generating the electron–hole pairs. It was reported in the research works that the device based on nano film, showed the enhancements of photocurrent were 10^3 to 10^6 times larger than the single nano-based structured device. This was due to the fact that photocurrent was calculated from many $NiCo_2O_4$ platelets than a single one. The response time of the nano film device was calculated through a 350 MHz Tektronix oscilloscope at 100 Hz and it was in the millisecond time scale for both rise and decay times which was shorter than any oxide photodetector. The device where a single nano structure was used, responded to the illuminated light based on the nano structure and was controlled by a complex process. But for nano film-based devices the quick response is controlled by the grain boundaries present in the structure. The nano films were pure and highly crystalline, which is also attributed to the efficiency of light absorption and the generation of photo carriers [39].

4 Perovskite Materials

With the advancement in the optoelectronic application, the need of ultra-broadband detection in the range of Ultraviolet (UV) to TeraHertz (THz) become highly essential as visible photodetection is not sufficient [40–42].

In this context, Perovskite material becomes promising for photodetection owing to high charge-carrier Mobility, broad light absorption, direct bandgap, long exciton diffusion length, and solution processability. The research interest in perovskite material has been enhanced after the discovery of perovskite-based solar cells in 2009 [43]. Over the years Perovskite material has shown photodetection mostly in the UV–Vis region [44–48]. So the researchers have taken different efforts to enhance the response wavelength beyond the Visual range either with the use of Sn instead of Pb [49] or by the addition of different Infrared (IR) absorption material [50–56]. Currently, the material is showing progress not only in photovoltaic cells [57–59], but also has attracted research interest in the enormous field like photodetectors [60, 61], lasers [62], and light-emitting diodes [63, 64].

Recently researchers prefer to fabricate devices with hybrid materials to get the advantage of stability and quick response of inorganic semiconductors and easy deposition technique, low cost & flexible fabrication of organic materials [65–67]. In this context, the hybrid lead halide perovskite-based device is unique and most promising having its direct, easy-tunable bandgap and solution processability [68]. The material is highly preferable due to the charge carriers owing to larger mobility, longer diffusion length, broader light absorption, and a longer lifetime [69–71]. The chemical formula is $CH_3NH_3PbX_3$ (Methylammonium lead halide perovskites) where X is either Iodine/Chlorine and/or Bromine. $CH_3NH_3PbI_3$ is mostly used among these for having a high lifetime and mobility of the charge carrier. The charge carriers are allowed to move far and the drawbacks of losing energy inside the cell are reduced.

A thin film of $CH_3NH_3PbX_3$ can be deposited generally through spin-coating. Stranks et al. reported that mixed halide ($CH_3NH_3PbI_3$-xClx) based nanostructured solar cells which shown 100 nm higher diffusion length than pure halide [43]. But the chemical instability of the Perovskite materials restricted the commercial viability of the materials [72]. To solve the issue Qasuria et al. fabricated UV and IR sensors from the degraded perovskite materials. This technology not only scavenges energy but also resolves lead contamination in the environment and reduces the cost and time for fabrication [73].

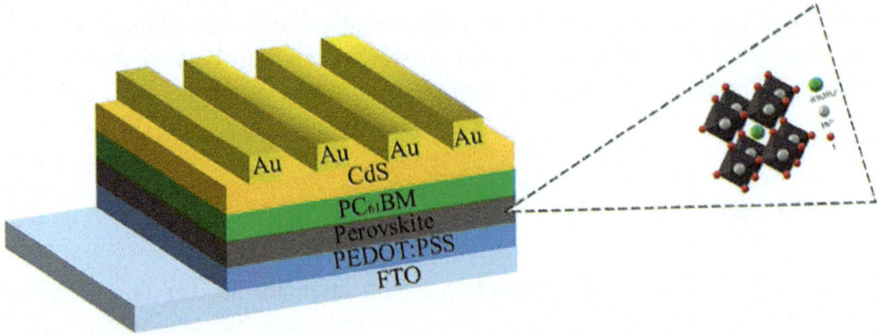

Fig. 14 Architecture of Perovskite based photodetector [73]

In this study the Perovskite film ($CH_3NH_3PbI_3$-xClx), the absorbent layer placed by the spin-coating method embedded betwixt two layers, one for Hole Transport and the other for Electron transport. For material preparation of the Hole Transport layer, the solution of PEDOT: Poly (3,4-ethylene dioxythiophene), and PSS: Poly (styrene sulfonate) have been accumulated with the spin coating technique over the FTO: Fluorine (F) doped tin oxide (SnO_2) coated glass. For material preparation of Electron transport, (6,6)-phenyl-C61-butyric acid methyl ester [$PC_{61}BM$] is used in addition to Cadmium Sulphide CdS. Gold electrode deposited on CdS. As a result, very promising performance was obtained. The capacitance and impedance have been investigated in the range of 0.01 to 200 kHz in different UV-IR intensities [73] (Fig. 14).

Li et al. reported a dual mechanism photodetector over the ultra-broadband in the range of Ultra-Violet (UV) to Terahertz (THz). The Photodetector was fabricated with the film of perovskite combined with organic and inorganic material ($CH_3NH_3PbI_3$). The photocurrent is generated in the range of UV–Vis with photoconductive effect and in the range of Near Infrared to Terahertz (NIR-THz) with bolometric effect. Significant responsivity, fast rise, and decay time were obtained in the UV-THz range [74].

References

1. H. Chen, K. Liu, L. Hu, A.A. Ghamdi, X. Fang, New concept of ultraviolet photodetectors. Mater. Today **18**, 493–502 (2015)
2. A. Aldalbahi, E. Li, M. Rivera, R. Velazquez, T. Altalhi, X.P. Peng Feng, A new approach for fabrications of SiC based photodetectors. (2016), www.nature.com/Scientificreports, doi:https://doi.org/10.1038/srep23457
3. X. Wang, Y. Zhang, X. Chen, M. He, C. Liu, Y. Yin, X. Zou, S. Li, Ultrafast, superhigh gain visible-blind UV detector and optical logic gates based on nonpolar a-axial GaN nanowire. Nanoscale. https://doi.org/10.1039/C4NR03581J
4. F. Gonzalez-Posada, R. Songmuang, M. Den Hertog, E. Monroy, Room-temperature photodetection dynamics of single GaN nanowires. Nano **12**, 172–176 (2012)

5. C. Soci, A. Zhang, B. Xiang, S.A. Dayeh, D.P.R. Aplin, J. Park, X.Y. Bao, Y.H. Lo, D. Wang, ZnO nanowire UV photodetectors with high internal gain. Nanoletters **7**, 1003–1009 (2007)
6. B.S. Zou, R. Liu, F. Wang, A. Pan, L. Cao, Z.L. Wang, Lasing mechanism of ZnO nanowires/nanobelts at room temperature. J PhysChem **B110**, 12865–12873 (2006)
7. U. Ozgur, Y.I. Alivov, C. Liu, A. Teke, M.A. Reshchikov, S. Doan, A.V.S.J. Cho, H. Morkocü, A comprehensive review of ZnO materials and devices. J. Appl. Phys. **98**, 041301 (2005)
8. R.H. Bube, *Photoelectronic properties of semiconductors* (Cambridge University Press, Cambridge, 1992)
9. J.S. Jie, W.J. Zhang, Y. Jiang, X.M. Meng, Y.Q. Li, S.T. Lee, Photo conductive characteristics of single crystal CdS nanoribbons. Nano Lett. **6**, 1887–1892 (2006)
10. W. Tian, et al., Flexible ultraviolet photodetectors with broad photoresponse based on branched ZnS-ZnOHeterostructureNanofilms. Adv. Mater. **26**, 3088–3093 (2004)
11. X. Huang, M. Wang, M.G. Willinger, L.D. Shao, D.S. Su, X.M. Meng, Assembly of three-dimensional hetero-epitaxial ZnO/ZnS core/shell nanorod and single crystalline hollow ZnS nanotube arrays. ACS Nano **6**, 7333 (2012)
12. L.F. Hu, J. Yan, M.Y. Liao, H.J. Xiang, X.G. Gong, L.D. Zhang, X.S. Fang, An optimized ultraviolet-a light photodetector with widerangephotoresponse based on ZnS/ZnO biaxial nanobelt. Adv. Mater. **24**, 2305–2309 (2012)
13. X. Fang et al., High-performance blue/ultraviolet-light-sensitive ZnSe-Nanobelt photodetectors. Adv. Mater. **21**, 5016–5021 (2009)
14. Z.W. Pan, Z.R. Dai, Z.L. Wang, Nanobelts of semiconducting oxides. Science **291**, 1947 (2001)
15. J.S. Jie, W.J. Zhang, Y. Jiang, X.M. Meng, Y.Q. Li, S.T. Lee, Photoconductive characteristics of single-crystal CdS nanoribbons. Nano Lett. **6**, 1887–1892 (2006)
16. X.H. Xie, Z.Z. Zhang, C.S. Shan, H.Y. Chen, D.Z. Shen, Dual-color ultraviolet photodetector based on mixed-phase-MgZnO/i-MgO/p-Si double Heterojunction. Appl. Phys. Lett. **101**, 081104 (2012)
17. K.S. Novoselov et al., Nat. Phys. **2**, 177 (2006)
18. F. Rana et al., Carrier recombination and generation rates for intravalley and intervalley phonon scattering in graphene. Phys. Rev. B **79**, 115447 (2009)
19. F. Wang et al., Gate-variable optical transitions in graphene. Science **320**, 206 (2008)
20. P.A. George et al., Ultrafast optical-pump terahertz-probe spectroscopy of the carrier relaxation and recombination dynamics in epitaxial graphene. Nano Lett. **8**, 4248–4251 (2008)
21. F. Xia, T. Mueller, Y.M. Lin, A.V. Garcia, P. Avouris, Ultrafast graphene photodetector. Nat. Nanotechnol. **4**, 839–843 (2009)
22. G. Konstantatos, M. Badioli, L. Gaudreau, J. Osmond, M. Bernechea, Hybrid graphene–quantum dot phototransistors with ultrahigh gain. Nat. Nanotechnol. **7**, 363–368 (2012)
23. F. Bonaccorso, Z. Sun, A.C. Ferrari, Graphene photonics and optoelectronics. Nat. Photon. **4**, 611–621. 2010
24. T. Gokus, R.R. Nair, A. Bonetti, M. Bohmler, A. Lombardo, K.S. Novoselov, A.K. Geim, A.C. Ferrari, A. Hartschuh, Making graphene luminescent by oxygen plasma treatment. ACS Nano **3**, 3963–3968
25. T.J. Echtermeyer, L. Britnell, P.K. Jasnos, A. Lombardo, R.V. Gorbachev, A.N. Grigorenko, A.K. Geim, A.C. Ferrari, K.S. Novoselov, Strong plasmonic enhancement of photovoltage in graphene. Nat. Commun. **2**, 458 (2011)
26. Y.H. Lee, X.Q. Zhang, W. Zhang, M.T. Chang, C.T. Lin, K.D. Chang, Y.C. Yu, J.T. Wang, C.S. Chang, L.J. Li et al., Synthesis of large-area MoS2 atomic layers with chemical vapor deposition. Adv. Mater. **24**, 2320–2325 (2012). https://doi.org/10.1002/adma.201104798
27. C.Y. Su, A.Y. Lu, C.Y. Wu, Y.T. Li, K.K. Liu, W. Zhang, S.Y. Lin, Z.Y. Juang, Y.L. Zhong, F.R. Chen, et al., Direct formation of wafer scale graphene thin layers on insulating substrates by chemical vapor deposition. Nano Lett. **11**, 3612–3616. https://doi.org/10.1021/nl201362n
28. D.S. Tsai, K.K. Liu, D.H. Lien, M.L. Tsai, C.F. Kang, C.A. Lin, L.J. Li, J.H. He, Few layer MoS2 with broadband high photogain and fast optical switching for use in harsh environments. ACS Nano **2013**(7), 3905–3911 (2013). https://doi.org/10.1021/nn305301b

29. Z. Yin, H. Li, H. Li, L. Jiang, Y. Shi, Y. Sun, G. Lu, Q. Zhang, X. Chen, H. Zhang, Single-layer MoS2 phototransistors. ACS Nano **6**, 74–80 (2012). https://doi.org/10.1021/nn2024557
30. G. Eda, H. Yamaguchi, D. Voiry, T. Fujita, M. Chen, M. Chhowalla, Photoluminescence from chemically exfoliated MoS2. Nano Lett. **11**, 5111–5116 (2011)
31. K.F. Mak, C. Lee, J. Hone, J. Shan, T.F. Heinz, Atomically thin MoS2: a new direct-gap semiconductor. Phys. Rev. lett. **105**, 136805 (2010)
32. H. Chen, L. Hu, X. Fang, L. Wu, General fabrication of monolayer SnO2 nanonets for high-performance ultraviolet photodetectors. Adv. Funct. Mater. **22**, 1229–1235 (2012). https://doi.org/10.1002/adfm.201102506
33. C. Li, H. Kan, J. Luo, C. Fu, J. Zhou, X. Liu, W. Wang, Q. Wei, Y. Fu, A high performance surface acoustic wave visible light sensor using novel materials: Bi2S3 nano-belts. RSC Adv. **10**, 8936–8940 (2020). https://doi.org/10.1039/c9ra08848b
34. B. Cui, H. Lin, Y.Z. Liu, J.B. Li, P. Sun, X.C. Zhao, C.J. Liu, Photo-physical and photo-catalytic properties of core-ring structured NiCo2O4 nano platelets. J. Phys. Chem. C **113**, 14083 (2009). https://doi.org/10.1021/jp900028t
35. Y. Kobayashi, X. Ke, H. Hata, P. Schiffer, T.E. Mallouk, Soft chemical conversion of layered double hydroxides to superparamagnetic spinel platelets. Chem. Mater. **20**, 2374–2381 (2008). https://doi.org/10.1021/Cm703443Q
36. L.F. Hu, R. Ma, T.C. Ozawa, F. Geng, N. Iyi, T. Sasaki, (2008) Oriented films of layered rare-earth hydroxide crystallites self-assembled at the hexane/water interface. Chem. Comm. **40**, 4897 (2008). https://doi.org/10.1039/B812111G
37. L.F. Hu, R. Ma, T.C. Ozawa, T. Sasaki, Oriented monolayer film of Gd2O3:0.05 Eu crystallites: quasi-topotactic transformation of the hydroxide film and drastic enhancement of photoluminescence properties. Angew. Chem. Int. Ed. **48**, 3846–3849 (2009). https://doi.org/10.1002/anie.200806206
38. L. Hu, L. Wu, M. Liao, X. Fang, High-performance NiCo2O4 nano film photo detectors fabricated by an interfacial self-assembly strategy. Adv. Mater. **23**, 1988–1992 (2011). https://doi.org/10.1002/adma.201004109
39. S. Mathur, S. Barth, H. Shen, J.C. Pyun, U. Werner, Size-dependent photo conductance in SnO2 nanowires. Small **1**, 713–717, (2005). https://doi.org/10.1002/smll.200400168
40. E.F. Schubert, Light-emitting diodes. Cambridge University Press, New York, 2006, pp. 35–40
41. D.Y. Kim, D.W. Song, N. Chopra, P. De Somer, F. So, Organic infrared upconversion device. Adv. Mater. **22**, 2260–2263 (2010)
42. V. Von Felbert, K. Kernland-Lang, G. Hoffmann, V. Wienert, D. Simon, T. Hunziker, Irradiation with water-filtered infrared A plus visible light improves cutaneous scleroderma lesions in a series of cases. Dermatology **222**, 347–357 (2011)
43. https://en.wikipedia.org/wiki/Perovskite_(structure)
44. M. Kim, G.-H. Kim, T.K. Lee, I.W. Choi, H.W. Choi, Y. Jo, Y.J. Yoon, J.W. Kim, J. Lee, D. Huh, H. Lee, S.K. Kwak, J.Y. Kim, D.S. Kim, Methylammonium chloride induces intermediate phase stabilization for efficient perovskite solar cells. Joule **3**, 2179–2192 (2019)
45. S.S. Shin, S.J. Lee, S.I. Seok, Metal oxide charge transport layers for efficient and stable perovskite solar cells. Adv. Funct. Mater. **29**, 1900455 (2019)
46. C. Xie, F. Yan, Enhanced performance of perovskite/organic-semiconductor hybrid heterojunction photodetectors with the electron trapping effects. J. Mater. Chem. C **6**, 1338–1342 (2018)
47. C. Xie, F. Yan, Perovskite/Poly(3-hexylthiophene)/graphene multihetero junction phototransistors with ultrahigh gain in broadband wavelength region. ACS Appl. Mater. Interfaces. **9**, 1569–1576 (2017)
48. M.I. Saidaminov, M.A. Haque, M. Savoie, A.L. Abdelhady, N. Cho, I. Dursun, U. Buttner, E. Alarousu, T. Wu, O.M. Bakr, Perovskite photodetectors operating in both narrowband and broadband regimes. Adv. Mater. **28**, 8144–8149 (2016)
49. M. Liu, Z. Chen, Q. Xue, S.H. Cheung, S.K. So, H.-L. Yip, Y. Cao, High performance low-bandgap perovskite solar cells based on a high quality mixed Sn–Pb perovskite film prepared by vacuum-assisted thermal annealing. J. Mater. Chem. Phys. **6**, 16347–16354 (2018)

50. F. Huang, Y. Peng, G. Liu, Toward ultrahigh sensitivity and UV–Vis–NIR broadband response of organolead halide perovskite/tin–phthalocyanine hetero structured photodetectors. J. Phys. Chem. C **123**, 11073–11080 (2019)
51. H. Sun, W. Tian, F. Cao, J. Xiong, L. Li, Ultrahigh-performance self-powered flexible double-twisted fibrous broadband perovskite photodetector. Adv. Mater. **30**, e1706986 (2018)
52. N. Alwadai, M.A. Haque, S. Mitra, T. Flemban, Y. Pak, T. Wu, I. Roqan, High-performance ultraviolet-to-infrared broadband perovskite photodetectors achieved via inter-/intraband transitions. ACS Appl. Mater. Interfaces. **9**, 37832–37838 (2017)
53. C. Xie, P. You, Z. Liu, L. Li, F. Yan, Ultrasensitive broadband phototransistors based on perovskite/organic-semiconductor vertical heterojunctions. Light Sci. Appl. **6**, e17023 (2017)
54. S. Chen, C. Teng, M. Zhang, Y. Li, D. Xie, G. Shi, A flexible UV-Vis-NIR photodetector based on a perovskite/conjugated-polymer composite. Adv. Mater. **28**, 5969–5974 (2016)
55. T. Gao, Q. Zhang, J. Chen, X. Xiong, T. Zhai, Performance-enhancing broadband and flexible photodetectors based on perovskite/ZnO-nanowire hybrid structures. Adv. Optic. Mater. **5**, 1700206 (2017)
56. S. Tong, J. Sun, C. Wang, Y. Huang, C. Zhang, J. Shen, H. Xie, D. Niu, S. Xiao, Y. Yuan, J. He, J. Yang, Y. Gao, High-performance broadband perovskite photodetectors based on CH $_3NH_3PbI_3$/C8BTBT heterojunction. Adv. Electron. Mater. **3**, 1700058 (2017)
57. M.L. Petrus, J. Schlipf, C. Li, T.P. Gujar, N. Giesbrecht, P. Müller-Buschbaum et al., Capturing the sun: a review of the challenges and perspectives of perovskite solar cells. Adv. Energy Mater. **7**, 1700264 (2017)
58. S. Mabrouk, M. Zhang, Z. Wang, M. Liang, B. Bahrami, Y. Wu et al., Dithieno [3, 2-b: 2', 3'-d] pyrrole-based hole transport materials for perovskite solar cells with efficiencies over18%. J. Mater. Chem. A. **6**, 7950–7958 (2018)
59. S. Mabrouk, A. Dubey, W. Zhang, N. Adhikari, B. Bahrami, M.N. Hasan et al., Increased efficiency for perovskite photovoltaics via doping the PbI2layer. J. PhysChem. C **120**, 24577–24582 (2016)
60. Y. Fang, Q. Dong, Y. Shao, Y. Yuan, J. Huang, Highly narrowband perovskite single-crystal photodetectors enabled by surface-charge recombination. Nat. Photon. **9**, 679 (2015)
61. H.-R. Xia, J. Li, W.-T. Sun, L.-M. Peng, Organohalideleadperovskite based photodetectors with much enhanced performance. ChemCommun. **50**, 13695–13697 (2014)
62. H. Zhu, Y. Fu, F. Meng, X. Wu, Z. Gong, Q. Ding et al., Lead halide perovskite nanowire lasers with low lasing thresholds and high-quality factors. Nat. Mater. **14**, 636–642 (2015)
63. M. Yuan, L.N. Quan, R. Comin, G. Walters, R. Sabatini, O. Voznyy et al., Perovskite energy funnels for efficient light-emitting diodes. Nat. Nanotechnol. **11**, 872–877 (2016)
64. X. Gong, Z. Yang, G. Walters, R. Comin, Z. Ning, E. Beauregard et al., Highly efficient quantum dot near-infrared light-emitting diodes. Nat. Photonics. **10**, 253–257 (2016)
65. M.D. Whitfield, S.S. Chan, R.B. Jackman, Thin film diamond photodiode for ultraviolet light detection. Appl. Phys. Lett. **68**, 290–292 (1996)
66. K. Liu, M. Sakurai, M. Aono, ZnO-based ultraviolet photodetectors. Sensors **10**, 8604–8634
67. K. Bayat, Y. Vygranenko, A. Sazonov, M. Farrokh-Baroughi, Design, fabrication and characterization of an a-Si: H-based UV detector for sunburn applications. SemicondSci. Technol. **21**, 1699–1702 (2006)
68. Y. Hou, E. Aydin, M.D. Bastiani, C. Xiao, F.H. Isikgor, D.J. Xue et al., Efficient tandem solar cells with solution-processed perovskite on textured crystalline silicon. Science **367**, 1135–1140 (2020)
69. W. Zhang, G.E. Eperon, H.J. Snaith, Metal halide perovskites for energy applications. Nat. Energy **1**, 1–8 (2016)
70. S.D. Stranks, H.J. Snaith, Metal-halide perovskites for photovoltaic and light-emitting devices. Nat. Nanotechnol. **10**, 391–402 (2015)
71. G. Kakavelakis, E. Gagaoudakis, K. Petridis, V. Petromichelaki, V. Binas, G. Kiriakidis et al., Solution processed $CH_3NH_3PbI_{3-x}Cl_x$ perovskite based self-powered ozone sensing element operated at room temperature. ACS Sens. **3**, 135–142 (2018)

72. K. Kwak, E. Lim, N. Ahn, J. Heo, K. Bang, S.K. Kim et al., An atomistic mechanism for the degradation of perovskite solar cells by trapped charge. Nanoscale **11**, 11369–11378 (2019)
73. T.A. Qasuria, N. Fatima, K.S. Karimov, A. Ibrahim, A novel and stable ultraviolet and infrared intensity sensor in impedance/capacitance modes fabricated from degraded $CH_3NH_3PbI_{3-x}Cl_x$ perovskite materials. J. Mater. Res. Technol. **9**(x x), 12795–12803 (2020)
74. Y. Li, Y. Zhang, T. Li, et al., Ultra-broadband, ultraviolet to terahertz and high sensitivity $CH_3NH_3PbI_3$ Perovskite photodetectors. ACS Paragon Plus Environ. (2020). https://doi.org/10.1021/acs.nanolett.0c00082

Optical Polarization Phase Modulation with Lithium Niobate

Ranjit Das and Rajib Chakraborty

1 Introduction

Lithium Niobate (LN, in short) is one of the most significant base elements for photonic integrated circuit design and plays the role as that of silicon (Si) for electronic ICs. LN single crystals and their periodically poled (PPLN) forms are important for various optical device fabrications such as for optical modulators, piezoelectric sensors, optical waveguides, sum/difference frequency generator, parametric oscillator, and for a variety of other linear and non-linear optical applications [1–4]. They are optically transparent and can transmit waves in the range 0.40–5.00 μm [5]. Normally at room temperature it behaves as ferroelectric material whereas above a critical temperature (Curie point, $T_C \approx 1190$ °C) it shows paraelectric properties [6].

Lithium niobate was first synthesized at Bell Telephone Laboratories (Murray Hill, New Jersey) by B. T. Matthias et al. in the year 1949 [7]. It has two varieties, the stoichiometric form, and congruent form and symbolized as SLN and CLN respectively. The two forms of the crystal are different in the sense that many of their physical characteristics and material/crystal parameters can differ by significant magnitude [8–12]. An important advantage of SLN over CLN is that the photo-refractive damage is reduced significantly in this crystal [13, 14].

A very striking combination of various optical features makes LN the most explored materials in recent years [1, 15]. This is a trigonal crystal of point group *3m* with negative uniaxial crystal axis and having an ordinary refractive index (2.297 at 633 nm) greater than the extraordinary index (2.208 at 633 nm), showing natural birefringence. The electro-optic effect in this crystal is much stronger than that in KDP or ADP. The various nonzero electro-optic coefficients of the crystal (CLN) are:

R. Das (✉)
Ramakrishna Mission Vidyamandira, Howrah, Belurmath, India
e-mail: ranjitdas.in@gmail.com

R. Chakraborty
University of Calcutta, Salt Lake, India

A. Saha et al. (eds.), *Optical to Terahertz Engineering*, Progress in Optical Science and Photonics 23, https://doi.org/10.1007/978-981-99-0228-6_6

$r_{13} = 10.0 \times 10^{-12}$ m/V, $r_{33} = 30.8 \times 10^{-12}$ m/V, [16, 17], $r_{22} = 3.4 \times 10^{-12}$ m/V and $r_{51} = 28.0 \times 10^{-12}$ m/V [1, 2].

The outstanding physical and optical properties of LN makes it the most desired material for optical modulators and integrated optical devices. The naturally birefringent LN crystal contains larger pyroelectric, piezoelectric, photo-elastic, acousto-optic, and electro-optic coefficients than its competitors. It also shows very strong bulk photorefractive and photovoltaic effects [8, 18]. Its significant Curie temperature (1100–1190 °C) makes it possible for the production of metal indiffused low-loss optical waveguides [19]. Again, it is thermally, mechanically and chemically stable. It is compatible with standard photonics and electronics integrated-circuit fabrication process. Its significant electro-optic coefficients are exploited in the infrared wavelength regime for Q-switching and optical modulation. Large nonlinear (d) coefficients of LN enables its use in the second harmonic generation, optical parametric oscillation and difference frequency generation and mixing for producing tunable infrared wavelengths [3, 5, 20].

A typical polarization phase tuning application based on optical polarization phase variation between the two orthogonal components (i.e., E-ray and O-ray) of the light wave is anticipated here. This application is built on the effect of electric field applied externally on LN crystal. It was observed that a trapezoidal-shaped LN crystal block can be more effectively employed to control the output intensity than by its regular geometrical forms [21]. A mathematical model has been proposed here for the trapezoidal-shaped LN crystal block to investigate the differential phase created between the two polarization components of light produced caused by electro-optic property of LN. Then the device is simulated with optimized parameters for desired phase change among the polarization components using genetic algorithm-based random optimization method. Again, using a multiple-strip structure of electrodes designed on an LN crystal block, the differential phase amongst the two polarization components is made to be tuned as can be seen here.

2 Tunable Differential Optical Polarization Phase Realization with LN Crystal

Various electrode geometries are in use in micro-fluidic systems and in integrated photonic applications for dielectrophoretic particle management, in-line scrutiny of biochemistry, and for controlling waveguides respectively [22–27]. Again, electrodes find applications in terahertz generation, phase control, sensing, etc. based on bulk crystal geometries [3, 28–30]. All these applications use electrode configurations, which are kept in parallel opposite faces and the associated analysis of the electric field is done by methods such as conformal mapping, Green's theorem, Fourier series expansion method, and many more [31–34]. Electric field analysis for out-of-plane electrodes sometimes needs to be addressed for studying gene electrotransfer, hybrid integrated circuits, and also for determining local field distribution in 2-dimensional space for in-vivo electro-chemotherapy [35, 36]. Bulk non-rectangular geometrical configurations are also in use for deflector-based switches [37]. Again, electro-optic device applications such as study of polarization phase in trapezoidal bulk crystal are addressed [38, 39]. It is reported that the sensitivity, to external field to be detected, can be enhanced with bulk geometry configuration [40]. But, the study of electric field for out-of-plane electrodes or inclined electrodes is not simple.

Here in this chapter the work is related with Schwarz-Christoffel Mapping (SCM), which is used as a technique for electric field distribution analysis in out-of-plane electrode-pair configuration. It is being a conformal mapping, has been applied for the study of fields for parallel electrode pairs [41–44]. Again, this technique find uses for the analysis of capacitance effects in non-parallel plate capacitors [44–46]. The same mapping is used here for analyzing the field for out-of-plane electrodes. Then, a variable or tunable phase is achieved on a trapezoidal-shaped LN crystal. In brief, an applied electric field (E) along the crystallographic axes and the linearly polarized light is fed through the z-direction of a trapezoidal-shaped LN single crystal block to realize a tunable differential optical polarization phase.

2.1 *Analysis of Electric Field by Schwarz-Christoffel Mapping or SCM Technique*

SCM is a conformal mapping alteration done on the upper half of the inner portion of a basic polygon. Here a set of complex numbers with only a positive imaginary part constitutes the upper half. The inner region of a polygon is mapped here from the Z-plane (known as physical plane) upon upper half section of the T-plane, which is also called the auxiliary plane. To evaluate the field, the non-uniform 2D field in the polygon is transformed into an equivalent rectangular section (parallel plate configuration), where the distribution of the electric field becomes uniform.

The electric field analysis inside the LN electro-optic phase tuning device, which is discussed in the following, can guide us to find the answer of an important question like, 'how does electric field modulate the light at every point, while it is passing through the device?' An in-depth analysis of electric field can also lead to improved performance of the electro-optic modulators and tuners.

Here, various analytical techniques are employed to evaluate the distribution of electric field for various configurations of electrode. First, a non-uniform field distribution study for parallel electrodes is discussed and then a non-uniform field distribution analysis for slant or inclined electrodes is considered.

Study of non-uniform field distribution in parallel electrodes. As mentioned in Ref. [47–49], the physical structure of parallel electrodes is assumed in the regime of complex Z-plane ($Z = x + iy$). Then, the right half of the section (colored red in Fig. 1a) is considered for the analysis. The coordinates of the associated rectangle are $Z_1 = \left(\frac{w}{2} + i\frac{h}{2}\right)$; $Z_2 = \left(i\frac{h}{2}\right)$; $Z_3 = \left(-i\frac{h}{2}\right)$, and $Z_4 = \left(\frac{w}{2} - i\frac{h}{2}\right)$; where h is the separation between the parallel electrodes and w is the width of the electrodes.

Then the walls of the channel are mapped along the real axis of t-plane (as shown in, Fig. 1b) by employing the following transformation equation given by [41]:

$$t = \sin\left(\pi \frac{iz}{h}\right) \tag{1}$$

Then the coordinates in t-plane are obtained by placing the coordinates Z_1, Z_2, Z_3, and Z_4 in Eq. (1). The midpoints of the Z_2 and Z_3 electrodes are placed at $\pm$ 1 points (that is at, t_2 and t_3, , respectively) on the t-plane real axis. Again, the standard equation of the SCM in the W-plane (shown in Fig. 1c) is given by,

$$w = \int_0^t \frac{k}{\sqrt{(1-u^2)(1-k^2u^2)}} du \tag{2}$$

where k is a constant, sometimes called the elliptical integral modulus of the first kind and is expressed as

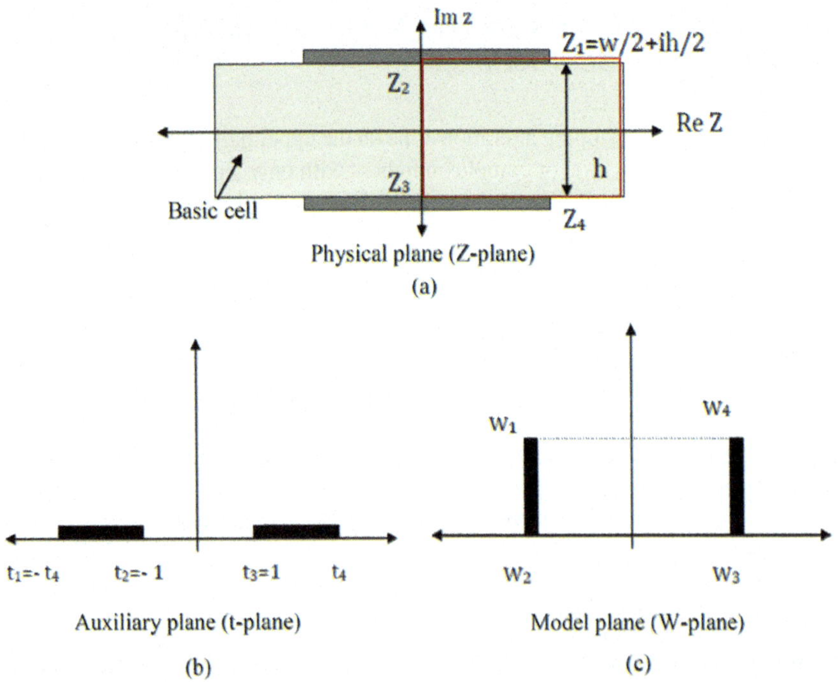

Fig. 1 **a** Cross-sectional view of parallel electrodes in Z-plane, **b** Z-plane to t-plane mapping, **c** t-plane to W-plane mapping [47]

$$k = \frac{1}{t_4} = \frac{1}{sin(iz_4/h)\pi} \tag{3}$$

The field distribution in the W-plane is hence given by the equation: $E_w = \frac{V}{2K(k^2)}$, where V is the applied voltage among the electrodes. So, the relationship between potential gradients in the Z-plane and W-plane is given by [41],

$$\nabla \emptyset_Z = \nabla \emptyset_W . \overline{f'(Z)} = \nabla \emptyset_W . \overline{\frac{dW}{dZ}} \tag{4}$$

The field distribution in the t-plane is obtained from Eq. (4) as: $E_T = E_W \frac{dW}{dt}$ So, the non-uniform field in the channel will be,

$$E_z = E_W . \overline{\frac{dW}{dt} \frac{dt}{dZ}}, \text{ or}$$

$$E_z = E_w \overline{k\left[\left(1 - t^2\right)\left(1 - k^2 t^2\right)\right]^{-1/2} \left[\left(\frac{i\pi}{h}\right)\cos\left(\frac{i\pi z}{h}\right)\right]} \tag{5}$$

Study of non-uniform field distribution with non-parallel electrodes. The SCM technique is also used here to evaluate the field distribution for the non-parallel or inclined electrode-pair configuration. At first, the cross-section of the inclined electrode-pair is mapped in Z-plane (Fig. 2a). From the Z-plane, the electrodes are then mapped in the t-plane and then consecutive mappings are followed in ς-plane and u-plane (Fig. 2b-d). These mappings are important for transferring the coordinate system from the non-parallel and unequal electrode regime to an equivalent parallel electrode configuration. Now one part of the electric field in Z-plane is kept between the two electrode planes, i.e., in the inner region subtended by the angle $\angle AOC (= \varnothing)$ and the remaining exists in the outer side the said area. For a very small angle of inclination ($\varnothing \to 0$) it is possible to approximate the electrodes are of nearly equal lengths and hence we can write $L_1 = L_2 = L$ and $\sin\varnothing = \varnothing$ (Fig. 2a). Also,

$$\left.\begin{aligned} r_1 &\approx \frac{h_1}{\emptyset} \text{and } r_2 \approx \frac{h_1 . \sin(\delta_1 - \emptyset)}{\emptyset}; \\ r_1 + L_1 &\approx \frac{h_2}{\emptyset} \text{and } r_2 + L_2 \approx \frac{h_2 . \sin(\delta_2 - \emptyset)}{\emptyset}; \end{aligned}\right\} \tag{6}$$

where $h_1 = [h - L\varnothing/2]$, $h_2 = [h + L\varnothing/2]$, $\delta_1 = [(\pi/2) - \varnothing]$, and $\delta_2 = [(\pi/2) + \varnothing]$. The angle, $\angle AOC$, being equal to 180° in t-plane (Fig. 2b), the transformation equation from Z-plane to t-plane is expressed as,

$$t = C_1 Z^{\pi/\emptyset} + C_0 \tag{7}$$

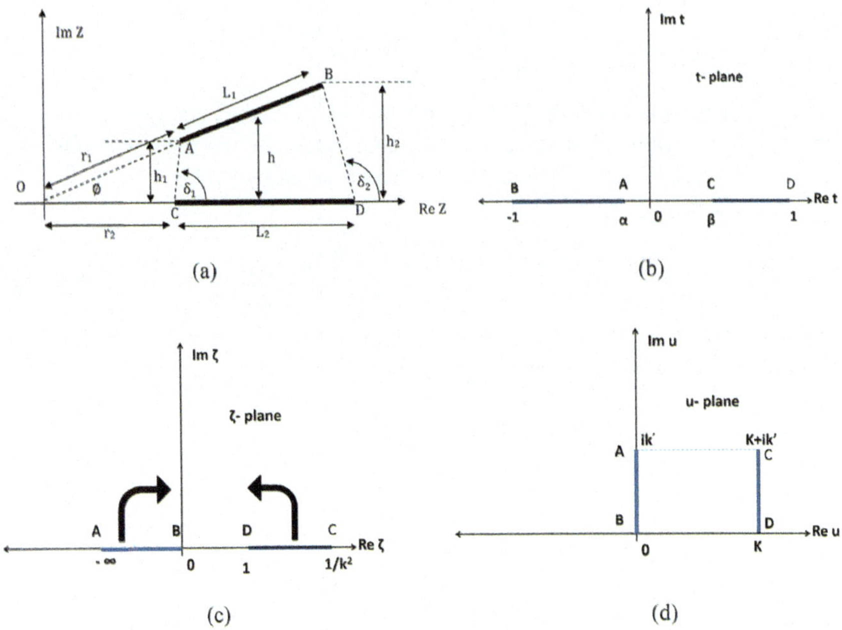

Fig. 2 **a** Inclined electrodes in Z-plane with an angle Ø, considering that the plates are extended at infinity **b** mapping in the t-plane **c** mapping in the ζ-plane **d** Schwarz- Christoffel Mapping in the u-plane [47]

So, coordinates in physical Z-plane may be obtained as: $Z_A = r_1.e^{i\varnothing}$, $Z_B = (r_1 + L_1).e^{i\varnothing}$, $Z_C = r_2$ and $Z_D = (r_2 + L_2)$. The Z-plane coordinates are then used for having the coordinates in t-plane so that the endpoints B and D of the electrodes are placed at -1 and $+1$ respectively and the other points A and C are mapped at $(\alpha, 0)$ and $(\beta, 0)$ respectively. So, the determination of the constants C_0 and C_1, is simplified. Hence, the coordinates at t and Z-plane are $t_B = -1$ and $Z_B = (r_1 + L_1).e^{i\varnothing}$ for B and the same are $t_D = +1$ and $Z_D = (r_2 + L_2)$ for D, respectively. Using these values in Eq. (7),

$$\left.\begin{aligned} C_0 &= \frac{(r_1 + L_1)^{\frac{\pi}{\varnothing}} - (r_2 + L_2)^{\frac{\pi}{\varnothing}}}{(r_1 + L_1)^{\frac{\pi}{\varnothing}} + (r_2 + L_2)^{\frac{\pi}{\varnothing}}} \\ \text{and, } C_1 &= \frac{2}{(r_1 + L_1)^{\frac{\pi}{\varnothing}} + (r_2 + L_2)^{\frac{\pi}{\varnothing}}} \end{aligned}\right\} \tag{8}$$

Substituting C_0 and C_1 in Eq. (7) the transformation equation becomes,

$$t = \frac{2Z^{\frac{\pi}{\varnothing}} + (r_1 + L_1)^{\frac{\pi}{\varnothing}} - (r_2 + L_2)^{\frac{\pi}{\varnothing}}}{(r_1 + L_1)^{\frac{\pi}{\varnothing}} + (r_2 + L_2)^{\frac{\pi}{\varnothing}}} \tag{9}$$

Using the coordinates of points A ($Z_A = r_1.e^{i\emptyset}, t = \alpha$) and C ($Z_C = r_2, t = \beta$) in t and Z-planes, two other coordinates α and β are then defined as,

$$\alpha = \frac{-2r_1^{\frac{\pi}{\emptyset}} + (r_1 + L_1)^{\frac{\pi}{\emptyset}} - (r_2 + L_2)^{\frac{\pi}{\emptyset}}}{(r_1 + L_1)^{\frac{\pi}{\emptyset}} + (r_2 + L_2)^{\frac{\pi}{\emptyset}}} \tag{10}$$

and

$$\beta = \frac{2r_2^{\frac{\pi}{\emptyset}} + (r_1 + L_1)^{\frac{\pi}{\emptyset}} - (r_2 + L_2)^{\frac{\pi}{\emptyset}}}{(r_1 + L_1)^{\frac{\pi}{\emptyset}} + (r_2 + L_2)^{\frac{\pi}{\emptyset}}} \tag{11}$$

Mapping in ζ-plane (Fig. 2c) is obtained in order to estimate the complex elliptical integral modulus of the first kind [45]. The transformation to ζ-plane from t-plane can be then expressed as,

$$\zeta = \frac{(1-\alpha).(1+t)}{2.(t-\alpha)} \tag{12}$$

The ς-plane transformation results in the coordinate distances which are suitable for SCM and consequently transfers the inner section of the electrodes to a rectangular region [46]. Replacing $t = \beta$ and $\varsigma = 1/\mathrm{k}^2$ in Eq. (12),

$$k = \left[\frac{2(\beta - \alpha)}{(1-\alpha)(1+\beta)}\right]^{1/2} \tag{13}$$

Then an upper half section of ς-plane is transferred to the inner portion of a rectangular area ABCD in u-plane (Fig. 2d). So, the SCM equation is given by,

$$\frac{du}{d\zeta} = A_1(\zeta)^{\frac{\pi}{2\pi}-1}.(\zeta - 1)^{\frac{\pi}{2\pi}-1}.\left(\zeta - \frac{1}{k^2}\right)^{\frac{\pi}{2\pi}-1} \tag{14}$$

Applying the coordinates of the points B ($\zeta_B = 0$) and D ($\zeta_D = 1$) in the ζ-plane, A_1 is obtained and is expressed as $A_1 = \frac{1}{k}$.

The coordinate transformation in Z, t, ζ, and u-plane are listed in Table 1 and given in the following.

The connection among the potential gradients in Z-plane and u-plane is then given by [42],

$$\nabla\emptyset_z = \nabla\emptyset_u.\overline{f'(Z)} = \nabla\emptyset_u.\overline{\frac{du}{dZ}} \tag{15}$$

The field pattern in u-plane is represented by $E_u = \frac{V}{K(k)}$. So, the field in Z-plane is obtained by using the Laplace equation and written as,

Table 1 Set of sequential coordinate transformations for non-parallel electrode [47]

Plane	Coordinates				Transformation
	A	B	C	D	
Z	$r_1 e^{i\Phi}$	$(r_1 + L_1)e^{i\Phi}$	r_2	$(r_2 + L_2)$	–
t	$(\alpha, 0)$	$(-1, 0)$	$(\beta, 0)$	$(1, 0)$	$t = C_1 Z^{\pi/\Phi} + C_0$
ς	$(-\infty, 0)$	$(0, 0)$	$(1/k^2, 0)$	$(-1, 0)$	$\varsigma = \frac{(1-\alpha)(1+t)}{2(t-\alpha)}$
u	iK'	0	$(K + iK')$	K	*Schwarz-Christoffel*

$$E_z = E_u \overline{\frac{du}{d\varsigma}\frac{d\varsigma}{dt}\frac{dt}{dz}} \tag{16}$$

Finally, the non-uniform field distribution associated with non-parallel or inclined electrode-pair is represented as,

$$E_z = -E_u \overline{\frac{\varsigma^{-1/2}(\varsigma - 1)^{-1/2}\left(\varsigma - \frac{1}{k^2}\right)^{-1/2}}{2k}}\,\overline{\frac{\left(1 - \alpha^2\right)}{2(t - \alpha)^2}}\,\overline{\frac{2.\pi.z^{\frac{\pi}{\emptyset}-1}}{\emptyset\left\{(r_1 + L_1)^{\frac{\pi}{\emptyset}} + (r_2 + L_2)^{\pi/\emptyset}\right\}}} \tag{17}$$

2.2 *Device Configuration with Non-parallel or Inclined Electrode Pair*

A device is projected here for this purpose with the dimensions and co-ordinate systems shown in Fig. 3. The optical axis of the bulk single LN crystal is assumed along the z-axis of the setup. Its one side is kept perpendicular to the z-axis while the opposite side is cut obliquely at an angle θ. So, a trapezoidal-shaped bulk crystal block of LN is obtained. When a narrow coherent beam of light is allowed to pass, along the y-axis, through the LN crystal block phase between the O-ray and the E-ray of polarization is varied as a function of the external electric field applied along the z-axis. Uniform metal electrodes are deposited on the slant surface and also on the opposite side of the structure. The inclined side is connected to the positive pole and opposite side with the negative pole of a constant DC source. Hence, the external field is applied along the z-axis while the input light wave is fed along the y-axis of the setup.

The shaded region in Fig. 4 denotes an arbitrary region of the electrodes. The electrodes, with respect to each other, are inclined at an angle θ. The distance between the electrodes is a function of height measured with respect to base of the structure. The gap between them is a function of L, h, and θ and is represented by $l(L, h, \theta)$. So, the field distribution along z-axis caused by the external voltage V is $E = V/l(L, h, \theta)$. For a given LN block, θ and L are constants. So, the resulting field (E)

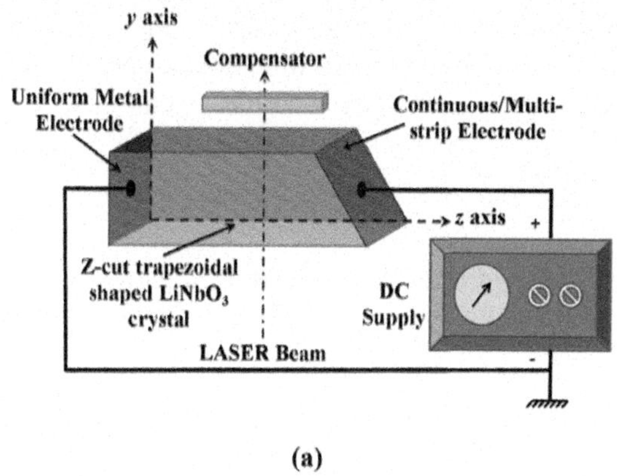

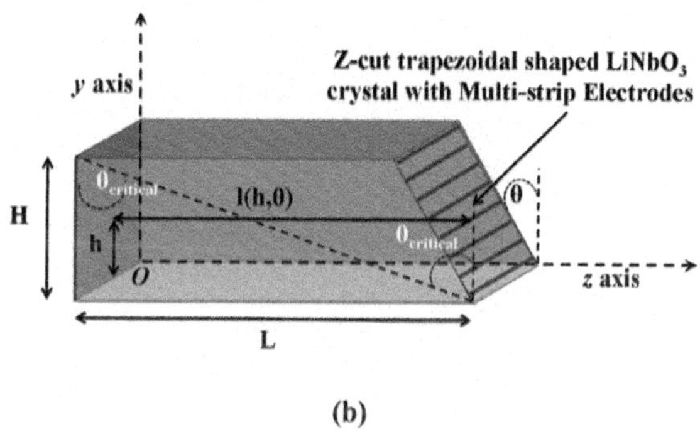

Fig. 3 **a** Schematic of the device with applied voltage, **b** the geometry of the proposed device with co-ordinate system used in the evaluation of electric field [47]

is a function of only height (h) of the considered electrode, for a constant voltage (V), and is represented by E_h, where,

$$E = E_h = \frac{V}{l(h)} = \frac{V}{L - h.\tan\theta} \tag{18}$$

This field E_h causes a change of phase of the polarization components due to a change in *r.i.* resulted from the electro-optic property of the crystal. The differential

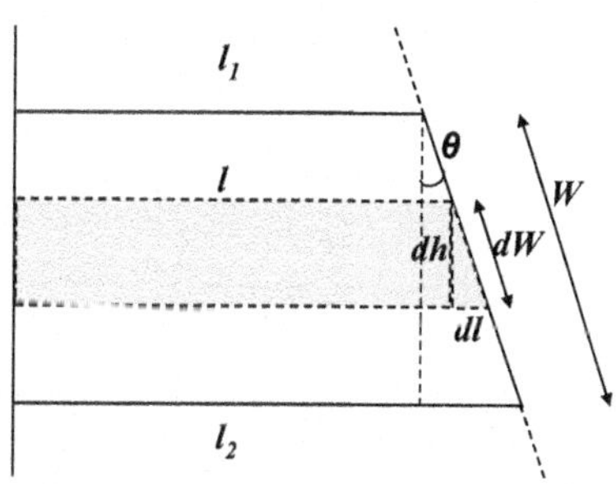

Fig. 4 An arbitrary strip with height h from the base of width W, slanted at an angle θ [47]

phase, however, will differ for the two polarization components along x axis and z axis. The equivalent phase difference among the two polarization components after travelling through a distance dh is given by,

$$d(\Delta\varphi) = \varphi_Z - \varphi_X = k_0(n_e - n_o)dh - \frac{1}{2}E_h k_0\left(n_e^3 r_{33} - n_o^3 r_{13}\right)dh \tag{19}$$

On the integration of the above function through the entire height (H) of the device, the first term in the above expression will give

$$\Delta\phi_1 = \int_O^H k_0(n_e - n_o)dh = k_0(n_e - n_o)H \tag{20}$$

as, $dl \cong 0$ for infinitesimal values of θ. This differential phase is caused by the intrinsic birefringence of LN and is closely controlled by the device height. So, using a suitable compensator this can be eliminated because it is fixed for any typical device. The second term in Eq. (19) is hence of importance and written as (Fig. 3),

$$d(\Delta\varphi_2) = \frac{1}{2}E_h k_0\left(n_e^3 r_{33} - n_o^3 r_{13}\right)dh$$

or,

$$d(\Delta\phi_2) = \frac{1}{2}\frac{V}{l}k_0\left(n_e^3 r_{33} - n_o^3 r_{13}\right)\cos\theta . dW = \frac{1}{2}\frac{V}{l}k_0\left(n_e^3 r_{33} - n_o^3 r_{13}\right)\cot\theta . dl \tag{21}$$

Clearly, this differential phase is caused by the electro-optic effect of LN. So, the lengths of the electrodes are important and must be used during analysis. Integrating the above expression in the limit $l = l_1$ to $l = l_2$, we get

$$\Delta\varphi_2 = \left[\frac{1}{2}V k_0\left(n_e^3 r_{33} - n_o^3 r_{13}\right)\cot\theta\right] . \int_{l_1}^{l_2}\frac{dl}{l}$$

or,

$$\Delta\varphi_2 = \left[\frac{1}{2}V k_0\left(n_e^3 r_{33} - n_o^3 r_{13}\right)\cot\theta\right] . \ln\left(\frac{l_2}{l_1}\right) = \left[\frac{1}{2}V k_0\left(n_e^3 r_{33} - n_o^3 r_{13}\right)\cot\theta\right] . \ln\left(\frac{l_2}{l_2 - W\sin\theta}\right) \tag{22}$$

The overall phase change ($\Delta\phi$) found from Eq. (19) contains two parts, $\Delta\phi_1$ and $\Delta\phi_2$. The $\Delta\phi_1$ is constant and may be compensated by using a compensator as mentioned earlier and schematically explained in Fig. 3a. Hence, the total phase change $\Delta\phi$ is entirely controlled by the change in phase $\Delta\phi_2$. Finally, l_2 is substituted by l, to obtain the general expression of differential phase among the two polarization components and is expressed as,

$$\Delta\varphi = \left[\frac{1}{2}Vk_0\left(n_e^3 r_{33} - n_o^3 r_{13}\right)\cot\theta\right].\ln\left(\frac{l}{l - W\sin\theta}\right) \tag{23}$$

It is obvious from Eq. (23), that for a given bulk LN crystal having continuous deposition of electrode pair placed with an angle θ between them, various amounts of differential phase can be obtained by choosing various magnitudes of l and/or W, even with a fixed applied voltage. So, it is possible to get various amounts of differential phase when the electrode is considered at different heights with or without varied electrode width. Again, tunability of the differential phase can be achieved by designing multiple-strip structure on the inclined face instead of the continuous electrode deposition. A spatial switch, specially designed for this purpose is so assumed that it may be exactly placed at any desired number of strip electrodes. So, the nth electrode (Fig. 4) may be projected to be the nth strip of the proposed multiple-strip configuration. Now, let us consider the base-to-base separation of the two nearest electrode strips is d. So, the nth strip's height with respect to the base of the bulk crystal is $d_n = (n-1)d$. Now, as shown in Fig. 5, we consider a multi-strip structure where l represents the length of the base of nth strip. So, we can write $l = L-(n-1)d.\tan\theta$.

So, the differential phase due to applied voltage (V) at n^{th} electrode strip, ($n = 1, 2, 3, \ldots\ldots, N$), is obtained from Eq. (23) and expressed as,

$$\Delta\phi_n = \left[\frac{1}{2}Vk_0\left(n_e^3 r_{33} - n_o^3 r_{13}\right)\cot\theta\right].\ln\left(\frac{L-(n-1)d\tan\theta}{L-(n-1)d\tan\theta - W\sin\theta}\right) \tag{24}$$

or,

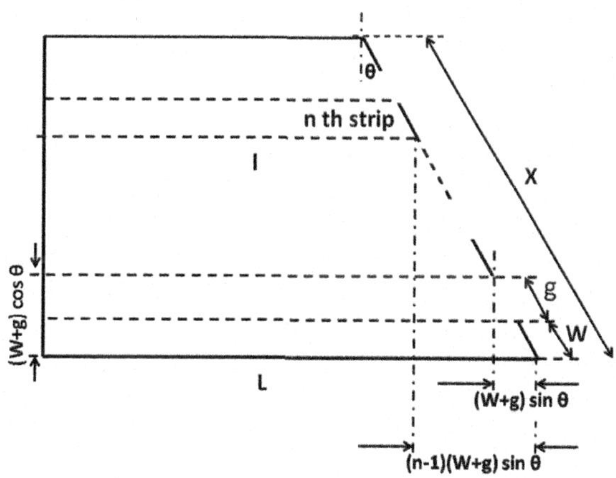

Fig. 5 Device geometry with multi-strip electrodes [47]

$$\Delta\phi_n = \left[\frac{1}{2}Vk_0\left(n_e^3 r_{33} - n_o^3 r_{13}\right)\cot\theta\right].\ln\left(\frac{L-(n-1)\frac{X}{n}\sin\theta}{L-[(n-1)\frac{X}{n}+W].\sin\theta}\right) \quad (25)$$

where N is the total number of strips on inclined face $= X/b$ (as, X = width of inclined face), $b = (W + g)$, (g and W are and distance among two successive stripes and the width respectively), and $d = (W + g)\cos\theta$. So, certainly the differential phase is lowest for the 1st strip and gradually enhanced with consecutive stripes. The differential phase among the polarization components also changes with the choice of a particular set of strip electrodes. So, by properly selecting the electrode strips, a tunable differential phase between the polarization components is achieved. This setup can then be efficiently applied for having tunable differential polarization-phase. Here, in this case Eq. (25) is expressed as:

$$\phi_n = \sum_1^N \left[\frac{1}{2}Vk_0\left(n_e^3 r_{33} - n_o^3 r_{13}\right)\cot\theta\right].\ln\left(\frac{L-(n-1)\frac{X}{n}\sin\theta}{L-[(n-1)\frac{X}{n}+W].\sin\theta}\right) \quad (26)$$

It is to be noted from Fig. 5 that at some critical angle of inclination the trapezoid will be transformed to a prism, and the above analysis is no more valid there. The critical angle of inclination can be shown to be,

$$\theta_{critical} = \cot^{-1}\left(\frac{H}{L}\right) \quad (27)$$

Hence, the inclination angle (θ) of the device can be changed from zero to an extreme value of $\theta_{Critical}$.

2.3 Distribution of Electric Field Inside the Device

In Fig. 6a the field pattern for a parallel plate structure for several channel heights is depicted. The electric field is clearly seen to be mainly concentrated near the surface of the device and the related electric fields are higher at the edges for the applied voltage of 500 V. By varying the channel height and/or by controlling the external voltage the interaction of the optical field with the electric field may be efficiently managed.

For a non-parallel electrode-pair configuration the field distribution is shown in Fig. 6b, which is of non-uniform polarization-phase variation. The evaluation of the expression of the electric field is associated with small angle approximation and hence, the electric field is plotted for an angle of inclination 8°. It is observed that the field displays unreliable characteristics for the angle of inclination above 12°.

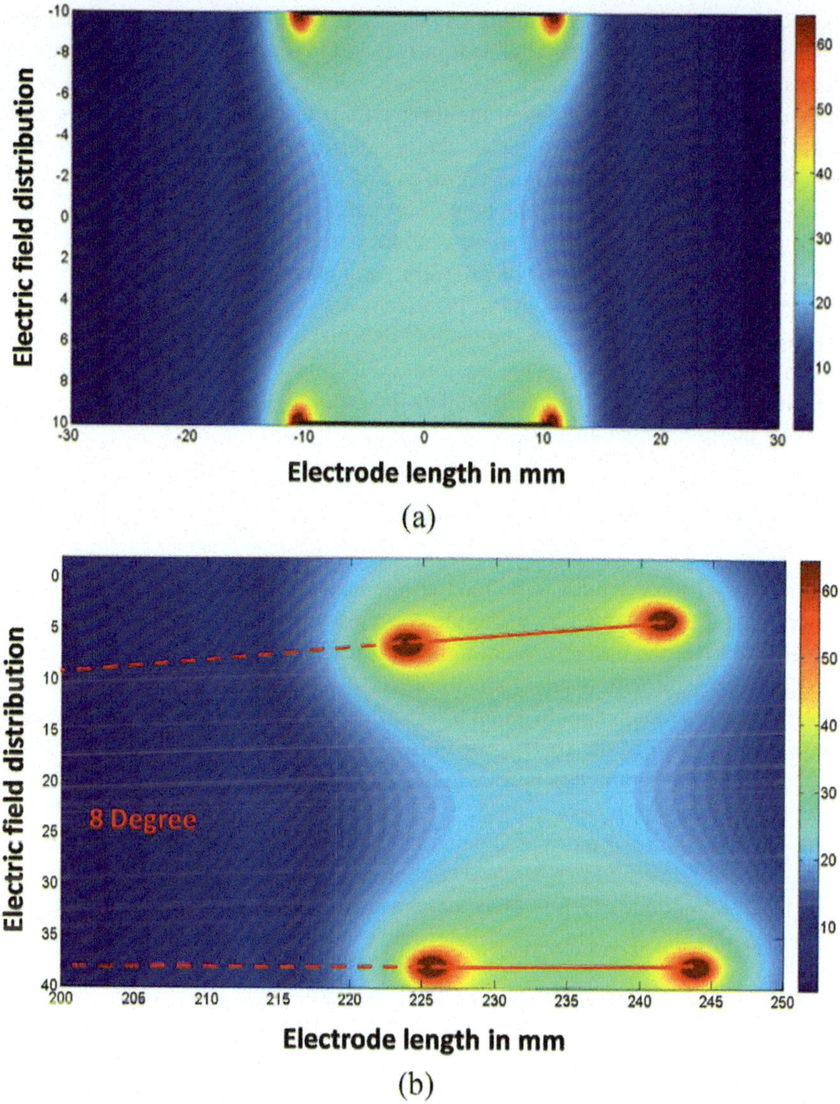

Fig. 6 Distribution of electric field for **a** parallel electrode geometry and for **b** inclined electrode pair geometry. Plots are obtained by MATLAB programming of electric field analysis using SCM [47]

2.4 *Optimization of Various Device Parameters*

The phase difference $\Delta\phi$ is a function of four parameters, which are applied voltage (V), the distance among the electrodes (l), the width (W) of the inclined electrode and inclination angle (θ) of the slanting plane. Though relation among the differential phase and the external voltage is linear, it is found that the extent of the differential phase cannot be defined beyond a typical magnitude of l and θ or for a lowest possible value of W. So, the dimensions of the three parameters, l, θ and W, do have some limits and they are to be optimized to obtain a desired differential phase. Moreover, any desired amount of differential phase can be obtained by choosing the suitable number of multiple strips with the help of a spatial switch.

To get a specific differential phase random optimization technique is used here for getting the optimum values of different parameters, which are *the width of the electrode* (W), *the distance between the electrodes* (l) and *the angle of inclination* (θ). Genetic algorithm based on the random optimization technique as given in [50], is employed here for designing the algorithm. First, the range of concerned variables with their accuracies are estimated and then the gene length for each variable is found by using the following equation,

$$gene\,length = ceiling\,function\left(\frac{\text{max value} - \text{min value}}{accuracy}\right) \tag{28}$$

The chromosome length is defined as the sum of various gene lengths. Then the gray string is generated by using the chromosome length. Single or multi-point transformation is the basis for random optimization. Logically, it signifies XOR*ing* which changes bit 0 into bit 1 and the reverse. Due to randomness in choosing between single-point and multi-point transformation or mutation, huge iterations are considered. After this, the transformed binary string is again converted to decimal. The scaling of the parameters is done to the desired range of this problem. The fittest outcome is achieved by comparison with the expected outcome using 1,000,000 iterations. The figure of merit associated with the optimization technique is defined by a parameter called *fitness function* (*FF*). It is also related to (1 + *merit function*) $^{-1}$. A larger magnitude of *FF* (~1) indicates a larger figure of merit, popularly known as *merit function* (*MF*). These parameters are related by,

$$MF = weight\ value(achieved\ value - targeted\ value)^2, \tag{29}$$

and,

$$FF = \frac{1}{1 + MF} \tag{30}$$

Furthermore, the algorithm is such that, the program becomes turned off when the fittest outcome (i.e. *FF* ~ 1) is attained, with the optimized values of the parameters.

In Tables 2 and 3, the optimized result to get quarter-wave and half-wave differential phase for a DC voltage of 500 V is given. It is observed that the optimized base length (L) is significantly low in comparison with the width of the electrode (W). So, the major interaction of light wave with the external electric field due to electro-optic property of LN is ensured. Due to the small angle approximation of the

Table 2 Device parameters with trapezoidal LN crystal for 90° or quater-wave differential phase [47]

Length (L) in mm	Width of electrode (W) in mm	Inclination angle (θ) in degree	Optimized phase (in degree)	Fitness Func. (FF)
5.1732	10.7638	14.4444	90.2486	0.9418
5.2520	8.3228	31.1111	89.9042	0.9909
1.6299	3.2835	16.1111	90.0330	0.9989
4.0709	6.8268	27.2222	89.9752	0.9994
2.4173	4.4646	21.1111	90.0204	0.9996
5.0945	10.9213	12.7778	90.0123	0.9998
4.4646	10.8425	6.6667	90.0013	1.0000
2.2598	5.4882	6.6667	90.0013	1.0000

Table 3 Device parameters with trapezoidal LN crystal for 180° or half-wave differential phase [47]

Length (L) in mm	Width of electrode (W) in mm	Inclination angle (θ) in degree	Optimized phase (in degree)	Fitness Func. (FF)
1.5512	3.2047	27.2222	179.1627	0.5879
1.3937	2.5748	31.6667	180.3476	0.8922
3.2835	9.3465	16.6667	179.8217	0.9692
4.2283	10.6063	20.5556	180.1057	0.9890
1.5512	8.3228	1.1111	179.9174	0.9932
2.6535	7.8504	15.5556	179.9203	0.9937
2.4173	8.7953	10	179.9782	0.9995
2.4173	10.8425	5	180.0000	1.0000

electrode modeling, the optimized inclination angle of the slanting plane is also very small and also shown in the Table.

2.5 *Introduction of Variable Phases*

For the said device with uniform continuous electrodes, variation of phase difference, $\Delta\phi$, with the change of either of the three independent parameters, the base length (L) of the trapezoidal crystal, the width (W) of the slant electrode and inclination angle (θ) of the slant electrode are analyzed and are shown in Fig. 7. The other parameter controlling the variation of phase difference is the applied voltage (V). However, there will be a linear dependence of phase difference with voltage and is

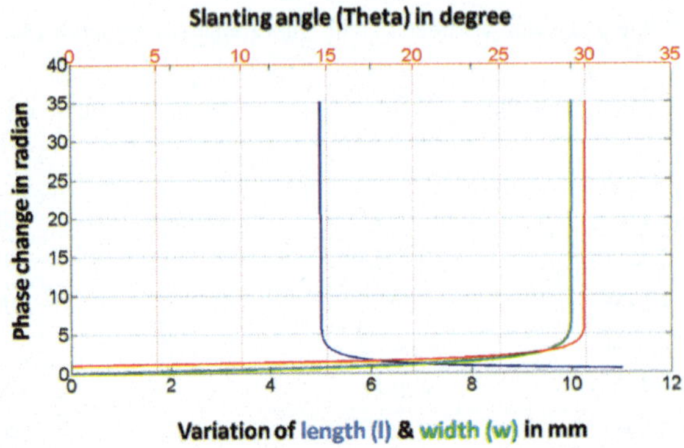

Fig. 7 Change of differential phase ($\Delta\phi$) between the polarization components of light wave **a** with the width, W of the inclined electrode strip and **b** with base length, l of the trapezoidal bulk crystal **c** with the angle of inclination, θ [47]

not plotted here. In plotting the data for variation of phase with any one parameter, the other two parameters are kept constant. Thus, keeping the base length fixed at 5 mm and inclination (θ) as 30°, the variation of phase ($\Delta\phi$) with slant electrode width (W) shows that the value of $\Delta\phi$ becomes imaginary beyond $W = 10$ mm (green coloured graph). This means that beyond a specific electrode width, the phase change is not possible while keeping the other two parameters fixed. Similarly, the phase variation becomes imaginary beyond a certain angle or if the base length is below a certain value (shown as red and blue colored lines respectively). In the above analysis the applied voltage is considered to be 500 V. The corresponding simulation is made by MATLAB.

2.6 *Tunability of the Device*

It is seen that tunability of the differential phase can be achieved by using multiple strip structures while joining the spatial switch to one or more desired strips. For having half-wave (180°) differential phase shift the optimized length of base (L) and width of electrode ($X = W$, for single continuous inclined electrode) are given in Table 3. Then, the variation of differential phase between the orthogonal polarization components with a change in strip electrode order number (i.e., the height of the concerned strip from the basement) is addressed. The constant ($W + g$), i.e., the sum of the width of a strip and distance between two consecutive strips is taken as 2.0 mm and the width of the slant electrode, X is assumed to be identical to the optimized electrode width of 10.84 mm for 180° differential phase. So, on the inclined face there will be 5 strips of electrodes and the differential phase shift for each electrode may be determined using Eq. (25). The differential phase changes associated with each electrode strip are then mentioned for four various combinations of W and g (Table 4). Set I of Table 4 shows the shift in phase for ($W + g = 2mm$) and it is found that the summation of all the phase shifts is nearly equal to 180°. Here we have used the values of electro-optic parameters for lithium niobate: $n_o = 2.297$,

Table 4 Differential phase change associated with different electrodes of the device for various sets of strip width (W) and channel gap (g) [47]

	Electrode no	Set I	Set II	Set III	Set IV
		$W = 2.0$ mm $g = 0.0$ mm	$W = 1.5$ mm $g = 0.5$ mm	$W = 1.0$ mm $g = 1.0$ mm	$W = 0.5$ mm $g = 1.5$ mm
Phase change (in degrees) associated with	1	27.3955	20.3520	13.4415	6.6591
	2	34.3727	25.4722	16.7830	8.2953
	3	37.5622	27.8042	18.2995	9.0354
	4	39.3900	29.1381	19.1654	9.4573
	5	40.5747	30.0017	19.7255	9.7299

$n_e = 2.208$, $r_{13} = 8.6 \times 10^{-12}$ m/V, $r_{33} = 30.8 \times 10^{-12}$ m/V. At $\theta_{Critical}$, the device geometry turns out to be a prism, as stated earlier. So, $\theta_{Critical}$ is the maximum allowable inclination or slanting angle. The value of differential phase ($\Delta\phi$) among the two components of polarization turns out to be indefinite and sign change $\Delta\phi$ is definitely not achievable as claimed in [39].

From the above discussion, it is clear that the angle (θ) has a significant impact on $\Delta\phi$, the phase among the two polarization components. The differential phase is a device geometry-dependent parameter and cannot be changed in a wide range. A different amount of typical differential phase is realized by placing the switch at the properly chosen strip electrode while the external DC voltage is kept constant. Connecting more and more electrodes to the spatial switch the magnitude of $\Delta\phi_{\max}$ may be further improved. So, the limit or range of tunability, i.e., ($\Delta\phi_{\max}- \Delta\phi_{\min}$) can be enhanced.

3 Conclusion

An application based on trapezoidal-shaped LN crystal block as a differential optical phase modulating device is proposed. For this, analyses of field distribution with parallel and inclined plane electrodes are made using the Schwarz-Christoffel Mapping method. But for non-parallel electrodes consistent result is only obtained by considering a small angle of inclination. These types of electrodes are simulated by using MATLAB® programming and the computational speed is tremendous. Further the non-parallel electrodes are considered to be placed on the slant surfaces of a trapezoidal block of LN bulk crystal. It is found that a differential polarization phase can be introduced by those electrodes by controlling various geometric parameters of the electrode structure. Optimized dimensional parameters of the device for a desired differential phase between the polarization components are achieved, and it is found that the tunability of this differential phase can be attained by considering multiple strip electrodes. This is really beneficial, as phase-tunability is achieved by taking only a constant DC voltage supply.

References

1. H. Nishihara, M. Haruna, T. Suhara, *Optical integrated circuit* (McGraw-Hill, UK, 1989)
2. A. Yariv, P. Yeh, *Optical waves in crystals: Wiley series in pure and applied optics* (John Wiley and Sons, New York, 1984)
3. K. Suizu, K. Kawase, Monochromatic-tunable terahertz-wave sources based on nonlinear frequency conversion using lithium niobate crystal. IEEE J. Sel. Top. Quantum Electron. **14**(2), 295–306 (2008). https://doi.org/10.1109/JSTQE.2007.911306
4. Y.Y. Lin, S.T. Lin, G.W. Chang, A.C. Chiang, Y.C. Huang, Y.H. Chen, Electro-optic periodically domain inverted lithium niobate Bragg modulator as a laser Q-switch. Opt. Lett. **32**(5), 545–547 (2007). https://doi.org/10.1364/OL.32.000545
5. INRAD Lithium Niobate. http://www.lambdaphoto.co.uk/pdfs/Inrad_datasheet_LNB.pdf
6. M.N. Palatnikov, N.V. Sidorov, V.I. Skiba, D.V. Makarov, I.V. Biryukova, Y.A. Serebryakov, O.E. Kravchenko, Y.I. Balabanov, V.T. Kalinnikov, Effects of nonstoichiometry and doping on the Curie temperature and defect structure of lithium niobate. Inorg. Mater. **36**(5), 489–493 (2000). https://doi.org/10.1007/BF02758054
7. B.T. Matthias, J. Remeika, Ferroelectricity in the ilmenite structure. Phys. Rev. **76**(12), 1886–1887 (1949). https://doi.org/10.1103/PhysRev.76.1886.2
8. A.M. Prokhorov, Y.S. Kuz'minov YS, *Physics and chemistry of crystalline lithium niobate.* (Adam Hilger, Bristol, 1990)
9. I. Pracka, A.L. Bajor, S.M. Kaczmarek, M. Wirkowicz, B. Kaczmarek, J. Kisielewski, T. Ukasiewicz, Growth and characterization of $LiNbO_3$ single crystals doped with Cu and Fe ions. Cryst. Res. Technol. **34**(5–6), 627–634 (1999). https://doi.org/10.1002/(SICI)1521-4079(199906)34:5/6%3c627::AID-CRAT627%3e3.0.CO;2-0
10. R. Mouras, M.D. Fontana, P. Bourson, A.V. Postnikov, Lattice site of Mg ion in $LiNbO_3$ crystal determined by Raman spectroscopy. J. Phys.: Condens. Matter **12**(23), 5053–5060 (2000). https://doi.org/10.1088/0953-8984/12/23/313
11. K. Kitamura, J.K. Yamamoto, N. Iyi, S. Kimura, T. Hayashi, Stoichiometric $LiNbO_3$ single crystal growth by double crucible Czochralski method using automatic power supply system. J. Cryst. Growth **116**(3–4), 327–332 (1992). https://doi.org/10.1016/0022-0248(92)90640-5
12. P.F. Bordui, R.G. Norwood, D.H. Hundt, M.M. Frejer, Preparation and characterization of off-congruent lithium niobate crystals. J. Appl. Phys. **71**(2), 875–879 (1992). https://doi.org/10.1063/1.351308
13. A. Ashkin, G.D. Boyd, J.M. Dziedzic, R.G. Smith, A.A. Ballman, J.J. Levinstein, K. Nassau, Optically induced refractive index inhomogeneities in $LiNbO_3$ and $LiTaO_3$. Appl. Phys. Lett. **9**(1), 72–73 (1966). https://doi.org/10.1063/1.1754607
14. D.A. Bryan, R. Gerson, H.E. Tomaschke, Increased optical damage resistance in lithium niobate. Appl. Phys. Lett. **44**(9), 847–849 (1984). https://doi.org/10.1063/1.94946
15. T. Volk, M. Wohlecke, *Lithium niobate: Defects, photorefraction and ferroelectric switching.* (Springer, Verlag, 2008)
16. M.D. Angelis, S.D. Nicola, A. Finizio, G. Pierattini, P. Ferraro, S. Grilli, M. Paturzo, Evaluation of the internal field in lithium niobate ferroelectric domains by an interferometric method. Appl. Phys. Lett. **85**(14), 2785–2787 (2004). https://doi.org/10.1063/1.1797534
17. Y. Chen, H. Zhan, B. Zhou, Refractive index modulation in periodically domain inverted MgO-doped congruent $LiNbO_3$ crystal. Appl. Phys. Lett. **93**(22), 222902 (2008). https://doi.org/10.1063/1.3042100
18. D.J. Paul, The progress towards terahertz quantum cascade lasers on silicon substrates. Laser & Photonics Reviews. **4**(5), 610–632 (2010)
19. R.V. Schmidt, I.P. Kaminow, Metal-diffused optical waveguides in $LiNbO_3$. Appl. Phys. Lett. **25**(8), 458–460 (1974). https://doi.org/10.1063/1.1655547
20. G.D. Miller, R.G. Batchko, W.M. Tulloch, D.R. Weise, M.M. Fejer, R.L. Byer, 42%-efficient single-pass CW second-harmonic generation in periodically poled lithium niobate. Opt. Lett. **22**(24), 1834–1836 (1997). https://doi.org/10.1364/OL.22.001834

21. S. Kim, V. Gopalan, K. Kitamura, Y. Furukawa, Domain reversal and nonstoichiometry in lithium tantalite. J. Appl. Phys. **90**(6), 2949–2963 (2001). https://doi.org/10.1063/1.1389525
22. R. Chakraborty, J.C. Biswas, S.K. Lahiri, Fabrication and characterization of ridge structures in $LiNbO_3$ for optical components. Opt. Eng. **43**(8), 1923–1926 (2004). https://doi.org/10.1117/1.1758270
23. G.M. Birnbaumer, P.A. Lieberzeit, L. Richter, R. Schirhag, M. Milnera, F.L. Dickert, A. Bailey, P. Ertl, Detection of viruses with molecularly imprinted polymers integrated on a microuidic biochip using contact-less dielectric microsensors. Lab Chip **9**(24), 3549–3556 (2009). https://doi.org/10.1039/B914738A
24. Z. Zou, J. Kai, M.J. Rust, J. Han, C.H. Ahn, Functionalized nano interdigitated electrodes arrays on polymer with integrated microuidics for direct bio-affinity sensing using impedimetric measurement. Sens. Actuators, A **136**(2), 518–526 (2007). https://doi.org/10.1016/j.sna.2006.12.006
25. N. Demierre, T. Braschler, P. Linderholm, U. Seger, H.V. Lintel, P. Renaud, Characterization and optimization of liquid electrodes for lateral Dielectrophoresis. Lab Chip **7**(3), 355–365 (2007). https://doi.org/10.1039/B612866A
26. D.F. Chen, H. Du, W.H. Li, Bioparticle separation and manipulation using dielectrophoresis. Sens. Actuators, A **133**(2), 329–334 (2007). https://doi.org/10.1016/j.sna.2006.06.029
27. H. Murata, Y. Okamura, High-speed signal processing utilizing polarization-reversed electro-optic devices. IEEE J. Lightwave Technol. **32**(20), 3403–3410 (2014). https://doi.org/10.1109/JLT.2014.2319454
28. D. Sando, E. Jaatinen, Optical data encryption using time-dependent dynamics of refractive index changes in $LiNbO_3$. Opt. Exp. **21**(17), 19510–19517 (2013). https://doi.org/10.1364/OE.21.019510
29. A. Garzarella, S.B. Qadri, T.J. Wieting, D.H. Wu, R.J. Hinton, Dielectrically induced sensitivity enhancements in electro-optic field sensors. Opt. Lett. **32**(8), 964–966 (2007). https://doi.org/10.1364/OL.32.000964
30. K. Liu, H.S. Kang, T.K. Kim, X.C. Zhang, Study of ZnCdTe crystals as terahertz wave emitters and detectors. Appl. Phys. Lett. **81**(22), 4115–4117 (2002). https://doi.org/10.1063/1.1524696
31. M. Gillick, I.D. Robertson, J.S. Joshi, Direct analytical solution for the electric field distribution at the conductor surfaces of coplanar waveguides. IEEE Trans. Microwave Theory Tech. **41**(1), 129–135 (1993). https://doi.org/10.1109/22.210239
32. E. Carlsson, S. Gevorgian, Conformal mapping of the field and charge distributions in multilayered substrate CPW's. IEEE Trans. Microwave Theory Tech. **47**(8), 1544–1552 (1999). https://doi.org/10.1109/22.780407
33. H. Morgan, A.G. Izquierdo, D. Bakewell, N.G. Green, A. Ramos, The dielectrophoretic and travelling wave forces generated by interdigitated electrode arrays: analytical solution using Fourier series. J. Phys. D: Appl. Phys. **34**(10), 1553–1561 (2001). https://doi.org/10.1088/0022-3727/34/10/316
34. X. Wang, X.B. Wang, F.F. Becker, P.R.C. Gascoyne, A theoretical method of electrical field analysis for dielectrophoretic electrode arrays using Green's theorem. J. Phys. D: Appl. Phys. **29**(6), 1649–1660 (1996). https://doi.org/10.1088/0022-3727/29/6/035
35. S. Corović, M. Pavlin, D. Miklavčič, Analytical and numerical quantification and comparison of the local electric field in the tissue for different electrode configurations. BioMed Eng Online **6**, 37 (2007). https://doi.org/10.1186/1475-925X-6-37
36. J.M. Bueno-Barrachina, C.S. Cañas-Peñuelas, S. Catalan-Izquierdo, F. Cavallé-Sesé, Capacitance evaluation on perpendicular plate capacitors by means of finite element analysis. in Intl Conf on Renewable Energies and Power Quality (ICREPQ'09). (Valencia, Spain, 2009), pp. 617–621. https://doi.org/10.24084/repqj07.452
37. Y. Zuo, M. Mony, B. Bahamin, E. Grondin, V. Aimez, D.V. Plant, Bulk electro-optic deflector-based switches. Appl. Opt. **46**(16), 3323–3331 (2007). https://doi.org/10.1364/AO.46.003323
38. P.E. Shames, P.C. Sun, Y. Fainman, Modeling of scattering and depolarizing electro-optic devices. II. Device simulation. Appl. Opt. **37**(17), 3726–3734 (1998). https://doi.org/10.1364/AO.37.003726

39. S. Sen, S. Mukhopadhyay, A noble technique of using a specially cut $LiNbO_3$ for achieving a greater amount phase difference between the components of light rays. Optik **124**(11), 1011–1013 (2013). https://doi.org/10.1016/j.ijleo.2013.01.021
40. A. Garzarella, D.H. Wu, Optimal crystal geometry and orientation in electric field sensing using electro-optic sensors. Opt. Lett. **37**(11), 2124–2126 (2012). https://doi.org/10.1364/OL.37.002124
41. R. Schinzinger, P.A.A. Laura, *Conformal mapping: Methods and applications* (Dover Publications Inc., New York, 2003)
42. T. Sun, N.G. Green, H. Morgan, Electric field analysis using Schwarz-Christoffel Mapping. J. Phys: Conf. Ser. **142**, 012029 (2008). https://doi.org/10.1088/1742-6596/142/1/012029
43. Markovic´ M, Jufer M, Perriard Y, Analyzing an electromechanical actuator by Schwarz-Christoffel mapping. IEEE Trans. Magnetics **40**(4), 1858–1863 (2004). https://doi.org/10.1109/TMAG.2004.828920
44. S.O.P. Blume, R. Ben-Mrad, P.E. Sullivan, Modelling the capacitance of multi-layer conductor facing interdigitated electrode structures. Sens. Actuators, B Chem. **213**, 423–433 (2015). https://doi.org/10.1016/j.snb.2015.02.088
45. Y. Xiang, The electrostatic capacitance of an inclined plate capacitor. J. Electrost **64**(1), 29–34 (2006). https://doi.org/10.1016/j.elstat.2005.05.002
46. B.R. Patla, Small angle approximation for non parallel plate capacitors with applications in experimental gravitation. Cornell University Library (2013). https://doi.org/10.48550/arXiv.1208.2984
47. R. Das, S. Ghosh, R. Chakraborty, Analysis of electric field for inclined electrodes and tunable differential polarization phase in a trapezoidal $LiNbO_3$ crystal. Eur. Phys. J. Appl. Phys. **72**(3), 30501(2014). https://doi.org/10.1051/epjap/2015150296
48. P. Linderholm, U. Seger, P. Renaud, Analytical expression for electric field between two facing strip electrodes in microchannel. Electron. Lett. **42**(3), 145–147 (2006). https://doi.org/10.1049/el:20063326
49. T. Sun, N.G. Green, S. Gawad, H. Morgan, Analytical electric field and sensitivity analysis for two microfluidic impedance cytometer designs. IET Nanobiotechnol. **1**(5), 69–79 (2007). https://doi.org/10.1049/iet-nbt:20070019
50. D. Beasley, D.R. Bull, R.R. Martin, An overview of genetic algorithm, Part 1. Fundamentals. Univ. Comput. **15**(2), 58–69 (1993)

Application of Fiber Optics for the Protection and Control of Power Systems

Nagendra Singh

1 Introduction

Electrical power systems, when viewed as being organized in hierarchical form, can be seen to have become complex in recent years due to their range of structural, status, and relevant technical issues. Such complexity, implemented in order to provide safer, cheaper, and more reliable power to the consumer, requires better control, monitoring, protection, measurement, and information transmission [1].

A prerequisite for the safe and stable operation of an electrical power system is the accurate and reliable measurement of the major system parameters, in particular current and voltage. In the Traditional power system, it is achieved by using a high-voltage current transformer and voltage transformers. But these transformers are very bulky in size and also very costly. Both these devices are increasingly coming under review in modem power systems due to their cost, and the safety implications for personnel, and the surrounding plants if a failure occurs. In the urban area cost of land and the cost of installation of a plant is high and also installation process is time-consuming. The main interests have been in producing devices to replace conventional equipment and CTs, utilizing the non-conductive nature of the optical components. Other important functions for measurements in the industry are those for diagnostic and metering tasks.

The main equipment used for the measurement of the quantities helps to solve the problems directly or indirectly arising in the power system. When a circuit breaker is used to break the current of the transmission line, due to the presence of the inductance and capacitance in the transmission line high-frequency surges can be induced. Managing the level of frequency and voltage is very important, in the power system for stable and safe operation.

N. Singh (✉)
Trinity College of Engineering and Technology Karimnagar, Karimnagar, Telangana, India
e-mail: nsingh007@rediffmail.com

A. Saha et al. (eds.), *Optical to Terahertz Engineering*, Progress in Optical Science and Photonics 23, https://doi.org/10.1007/978-981-99-0228-6_7

Since the power system required stable and efficient operation hence the transmission of data is very complex. For the stable operation and control of the power system, the centralized control system was used which required high bandwidth of the communication system. But the traditional control and communication system has many limitations in respect of frequency allocations and channel availability [2].

The classical control systems are also very difficult to handle and operate because of their complex nature and installation cost and place requirement is very high. Since a power system uses high voltage and current for transmission so its needs a reliable control system. And the control system depends on the signals it received from various parts of the operating area of the power system. If they received information very fast then they can take quick action and operate the system and hence the system will be safe and operated as per desired operations.

Old communication channels used metallic circuits, which can be affected by external causes like interference of voltage and current from other circuits. Hence the effectiveness and reliability of such a metallic-based control system are reduced. Mostly pilot circuits are used for communication of the signal which is purchased or rented by telecommunication authorities. Such kind of communication circuits used electronic signal repeaters which are unable to send the signals long distances. Since the power system used a large network for long-distance its need a suitable system that can send the signal in all operating area at high speed without any interference [3].

The main objective of the use of the pilot wire is to trip the circuit when any unwanted operation arises in the power system. Since the operation of the pilot wire depends on the signal of CT and PT, so it can help to protect the system if received correct and fast signals from CT and PT.

Continuous monitoring and maintenance are required for the equipment and systems used for measurements and control of the electrical power systems because they are sensitive and very costly. Now we are required to use online monitoring approach which is effective, accurate, and less costly as compared to traditional systems. Such online monitoring helps to operate the system with stable voltage, frequency, and power factor and also helps to protect the whole power system from unwanted hazards. Since the traditional monitoring and control systems are not much efficient so cannot perform effectively the expected tasks.

For the support of the monitoring and control system, the required to stabilize a set of locations (monitoring point) in the electrical power system, so the system can measure the accurate value and also communicate with high speed the same exact values. Such kind of development of system improves the working efficiency of the control system. Integration and coordination of the operation of a utility system based on a hierarchical control superimposed on a distributed local control structure require local microcomputer networks linked together and installed in key control areas such as power plants, substations, and load centers, for example.

Modern power systems use SCADA (supervisory control and data acquisition) monitoring and control system which is operated by the local network system. Vibration monitoring, control of large remote heat systems, remote control of transformers, temperature monitoring of high-voltage cables, and lightning observations from other functions of monitoring and control are also important. The fault is a common problem in transmission lines, especially in rainy sessions, which interrupt the supply of power for a long time [4].

Optical fiber transmission systems are very fast and free from external interference used in the electrical power system for control and operation. In these respects optical Fiber thermometer can offer a new perspective. OFC transmission system is used between substation for transmission line safety, monitoring between point-to-point working systems with connection to the SCADA system, and transmitting voice or video messages for supervision and control. What is more, to support the complicated system management and improved protection algorithms of complex structures of substation protection and control systems, integrated digital protection and control systems have been employed.

Electrical equipment that fulfills responsible functions in high voltage, high power systems should be supervised continuously in the real–time of operation. This equipment includes power lines, transformers, switches, current transformers, etc. The supervision functions and system redundancy is needed to assure the desired level of reliability in the system [5].

The protection system mainly divided in two-level, higher level, and low level. The higher level consists of the diagnosis of the technical problems arises in the system and lower level included buzzer/alarms and operating switches. Technical diagnostics sub–system gives the possibility of early or nearing fault detection to the user of the whole system. The build feature enabling an early detection in the system protects it directly against any fatal breakdown stage. Here system used for supervision and control is distinguished in four following reliability stages:

- Normal work under nominal conditions,
- Early fault stage of work,
- Disturbed stage of work,
- Break down stage.

The electrical power system is supervised continuously so that power can be transmitted and distributed at all stages of the system without any interruptions. During the breakdown of power, information received by the supervision system is very important. This information is shared in many high and down-level systems to overcome the breakdown problem as soon as possible. The optical fiber communication system here helps to transmit the information very quickly to supervision control units as well as other stages without any change in signals.

This chapter focused on the fiber optic supervision network system connected with SCADA supervisory system and protects the whole power system. The supervision system mainly depends on the signal received in terms of voltage, current, frequency, temperature, power factor, and change in lodging at receiving end side. Also using many electrical switches and equipment which is used to trace a double in the electrical power system and also trip the circuit when any unwanted state arises with the help of a supervisory system [6].

2 Optical Fiber System

Glass or fiber is used to make the optical fiber. Optical fiber is a very thin wire and its diameter is near to diameter of a human hair. It is a very long wire where the signal is transmitted at almost the speed of light using the principle of total refraction from the center of the wire from one end to another end [7]. For many applications, it performs better than metallic conductors because it's had higher bandwidth and transmits speeds. A comparative analysis between copper conductor communication and OFC is shown in Table 1.

Table 1 Comparison between OFC and copper conductor communication

Properties	OFC	Copper conductor
Speed	OFC can transmit signal with speed of light	Its speed is less than OFC
Electrical Isolation	Free from ground loop problem	Ground compulsory required
Sparking	No sparking	Maybe sparking arises
Interference	No electromagnetic interference	electromagnetic interference arises
Power loss	Very low	High
Handling	Its weight is very light so easily can be handled	Its weight is high so transportation and handling are not easy

2.1 *Construction of Optical Fiber*

A Fiber Optic Cable consists of the following parts shown in Fig. 1.

A. **Core**: It is in cylindrical form made of plastic used to cover the fiber cable along with the length of the cable. The core provides protection for the cable. The length and diameter of the core design are as per the applications used.
B. **Cladding**: It is used for the protection of the core. It covers the core outer part. It helps the light to reflect back into the core. Generally, when light enters from dense medium (core) to low dense medium (cladding), it changes its angle in such a way that it can reflect back to the core.
C. **Buffer**: Since the fibers are arranged in thousands of optical fibers so its required mechanical support and protection, such mechanical support, and protection are provided by the buffer.
D. **Jacket:** It is the outer part of the fiber cable used for the protection of the cables from outside hazards and environmental problems.

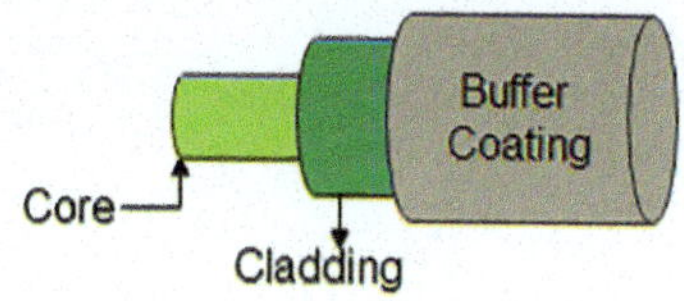

Fig. 1 Main part used in optical fiber

2.2 *OFC Operating Principal*

Characteristic of any Optical materials depends on the index of refraction as they are reflected in the medium of propagation changed. The optical fiber is working on the principle of total internal reflections as shown in Fig. 2. The reflection index is defined as

$$\text{Refraction index} = \frac{\text{speed of light in a vacuum}}{\text{speed of light in the material}}$$

Law of refraction given by Eq. (1)

$$n_I sinI = n_r sinR \tag{1}$$

where I-The angle of incidence, R-angle of reflection, n_I -medium of incidence of the light, and n_R–medium of reflection of light.

For the total internal reflection, it is required to angle of incidence is always greater than the critical angle. In the optical fiber, about 82° incidence angle is used for internal reflection. At such an angle, light is reflected without any loss as shown in Fig. 3.

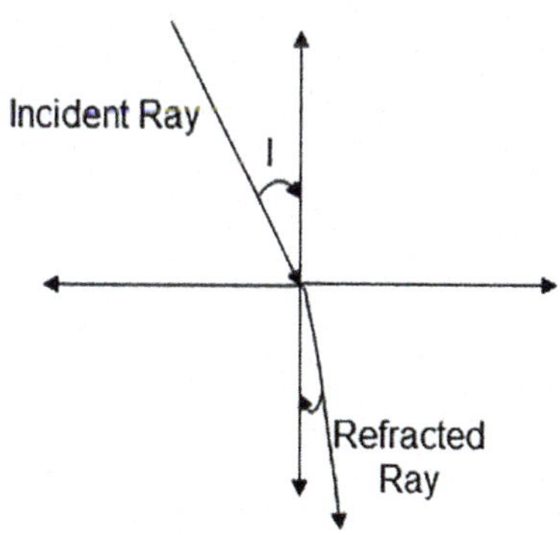

Fig. 2 Principle of interference used in optical fiber

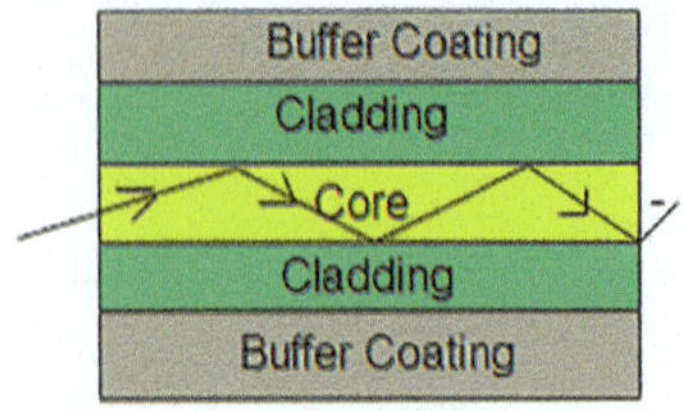

Fig. 3 Total internal reflection of light inside of the core of the fiber optics

2.3 *Classifications of Optical Fibers*

As per the construction and used material composition for core optical fiber cables are categorized as shown in Fig. 4.

1. **Step-index fiber**—In this case of the cabal, the core has a uniform refractive index throughout the length and it is suddenly changed when reached the cladding boundary.
2. **Graded-index fiber**—In this case of the cabal, the core has a variable refractive index from the center of the cable.

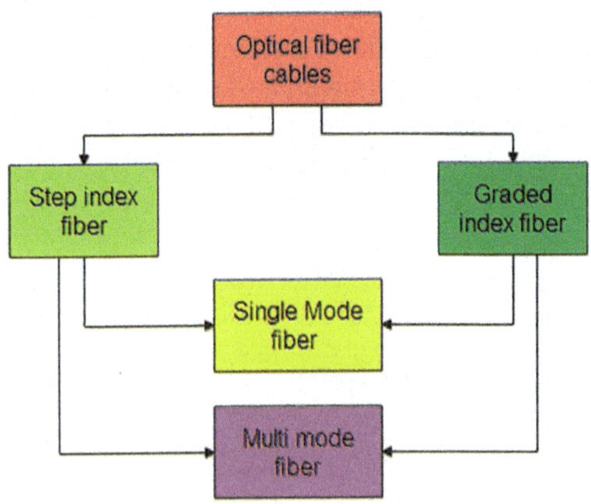

Fig. 4 Classifications of the optical fiber cables

Now these two types of cables are classified into two forms.

A. **Single-mode fiber**

- It is suitable for long-distance communication only.
- Small diameter of the glass used for the construction of single-mode fiber.
- Minor signal strength reduced due to small diameter
- Signal light of beam used which can travel for long distance.
- High bandwidth used to transmit the signal
- In this mode laser is used as a source of light

B. **Multi-mode fiber**

- It is used for a short-distance of communication.
- It required a large core diameter for the reflection of the signals.
- Multi signals can be transmitted
- Signal loss is possible in multi-mode fiber
- LED used as a source of light

2.4 Advantages of Fiber Optics

OFC has the following advantages

- OFC can support higher bandwidth communication.
- Signals travel for long distances so need for any signal booster.
- OFC signal cannot affect by electromagnetic interference.
- It can be used in water also.
- OFC has light in weight; mechanical strength can be used for long life, and easy in handling.
- It is free from maintenance.

2.5 Disadvantages of OFC

- It is costly compared to copper wire.
- Since optical fiber uses glass so it required higher protection compared to copper wire.
- New Installation required much labor and hence needs much cost.
- If the cable bend or curved around a radius of a few centimeters due to its fragile nature it can be broken or the signal can be lost.

2.6 Fiber Optic Communication

It is a technology used to transmit the data from one end to another end of the circuit. In this system, infrared light pulses are used as signals to transmit the data. Light is used as a signal carrier and it is converted into data. Where long distance, high bandwidth, and high-speed data transfer are required optical fiber cable is preferred instead of copper cable [8].

For the transmission of audio, video, and large data Fiber-optic communication is the most suitable system. Optical fiber communication is also used for telephone and television signals. Currently in India Jio and Airtel companies mostly use fiber optical systems for internet and TV signal communication.

2.6.1 Working of Fiber Optic Communication System

In an optical fiber, the light signal is transmitted from one point to another end. As shown in Fig. 5 it has a transmitter that transmits the electrical signal to the light of the source and such electrical is converted into a light signal. After that using fiber and optical cables light source signal send to the receiver, and here again, it is converted into a digital signal for further use [9].

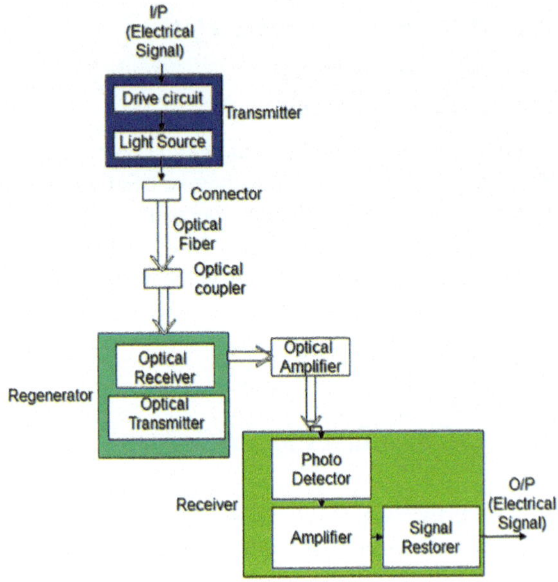

Fig. 5 Main part used in communication

2.6.2 Elements of a Fiber Optic Communication System

Figure 5 shows the basic major elements used for the optical fiber communication system.

Transmitter—It consists of a drive circuit and light sources. A drive circuit is generally used to convert the electrical signal into a digital signal. And a light source has an LED and laser diode for the generation of the light beam. The difference between LED and laser light diodes is given in Table 2. Generally, the light source value depends on the network area and size of the network.

LED is an important element used as a light source In an optical fiber communication system. It can convert the electrical signal into an equivalent light signal which is used as an input signal for optical fiber. The LED has some good characteristics like it induces less heat when converting an electrical signal into a light

Table 2 Comparison of characteristics of laser diode and LED

Quantity	Characteristic of laser	Characteristic of LED
Speed	Very fast	Slow
Handling	Difficult compared to LED	Easy in operation
Output power	High	Low
Cost	High	Low
Spectral width	Narrow	High

source so it improves the overall efficiency of communication. LED has the following advantages:-

- Size is small so it easily fits in the circuit.
- LED has a high radiance.
- LED has small emitting regions like optical fiber dimensions so it is suitable for use.
- LED is a highly reliable element.
- They can be changed at high speeds.

Optical fiber cables: it is used to transmit electrical signal in terms of the light beam from one circuit to another circuit. The main characteristic of such optical fiber is high bandwidth, Low power losses, and high speed of transmitting the data.

Regenerator: It consists of an optical receiver and optical transmitter. The optical receiver receives the optical signal coming from the optical fiber and the optical transmitter transmits the optical signal to the receiver via an optical amplifier.

Receiver: It consists of the photodetector, which identifies the signal that came from the regenerator. A signal restorer is also present in the receiver, which converts the light signal into an electrical signal.

2.6.3 Characteristics of Optical Fiber

Optical fiber has many characteristics as shown in Fig. 6.

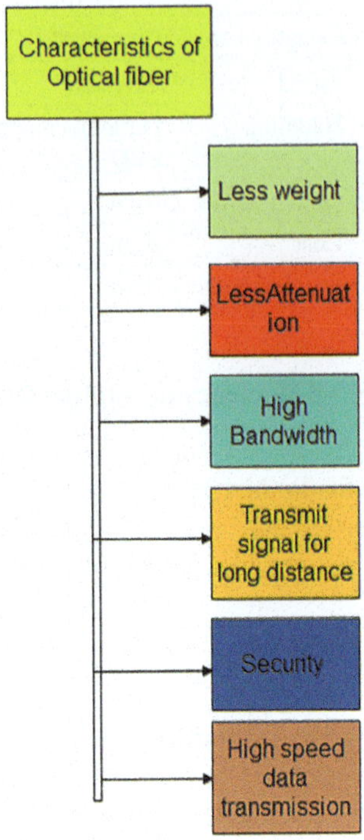

Fig. 6 Characteristic of optical fiber

2.6.4 Fiber Optics Applications

Optical fibers have many applications, and Fig. 7 shows various sectors where optical fiber is highly recommended [10].

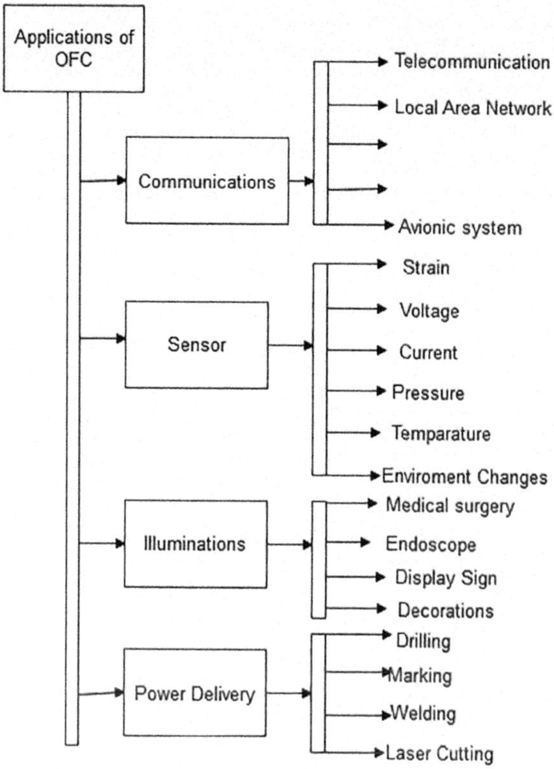

Fig. 7 Applications of OFC

2.6.5 Fiber Optic and Satellite Communication

In the present day, Optic fiber and satellite communications are used mostly all over the world. Both technologies have some advantages and limitations. The main differences between these methods are shown in Table 3.

Table 3 Difference between fiber optical and satellite communication

S.no	Fiber optic communication	Satellite communication
1	It is more suitable for urban areas	It is suitable for anywhere in the country
2	This type of communication used light signals for the medium of propagation	Electromagnetic signals are used for the medium of propagation
3	For this communication optical fiber is required	In this communication relay station used
4	Bandwidth of the fiber optical communication is very high and low interference with	Bandwidth is less and interference is always flat
5	For fiber communication antennas are not required	Special antennas required
6	Medium of transmission is fiber	Medium of transmission is air
7	It is more suitable for point-to-point and short-distance communication	It is suitable for long-distance communication only
8	Rate of data flow is very high and propagation delays are very low	Flow of data rate is slow and since data is transferred from satellites to earth, propagation delay is high
9	Its cost is less	It is very costly
10	It is used for stationary communication	It can be used for movable communication as the mobile
11	It is reliable and chances of error are very less	It is also reliable but the chances of error are more

Advantages

The advantages of optical fiber communication include the following.

- Communication is secured
- Speed
- Electromagnetic compatibility
- Distance
- Bandwidth
- No Power loss
- Interference
- Size
- Less Weight
- Security

3 Electrical Power System Protection and Control System

3.1 Power System Protection

The electrical power system is designed in such a way that it can generate and transmit power to all consumers. Power systems ensure operation economically and provide full safety to the consumers. The huge capital investments in the generation, transmission, and distribution of power with proper precautions are taken to ensure that the equipment operates at peak efficiency and must be protected from accidents. The electrical current is normally transferred from the generator to the transformer, transformer to transmission lines, transmission lines to the distribution line, and distribution line to different loads using suitable insulators [1].

Total protection depends on the insulation provided in the power system, so if due to environmental effects or aging or any physical accident insulator is damaged then the current follows an abnormal path generally known as a short circuit (fault). Generally, we can say fault means the abnormal operation of the power system. Faults may arise in two ways, first open circuits and second short circuits. Various faults occurred in the transmission line shown in Fig. 8. In most cases, short circuit faults arise more compared to open circuit faults [11].

During the fault occurs in the power system, an automatic protective device is required to isolate the faulty element as quickly as possible to keep the healthy section of the system in normal operation. When any short circuit fault occurs in any section of the power system, a very high short circuit current will start to flow in the circuit this high current is very dangerous for the human as well as equipment connected with the power system. In such conditions, it is required to clear the fault within a fraction of a second. If the short circuit is present in the circuit for a long time it can damage the system mostly and sometimes it catches the fire and damages the surrounding areas. So it is required to eliminate short circuits as quickly as possible using protective elements like relays and circuit breakers. Also required

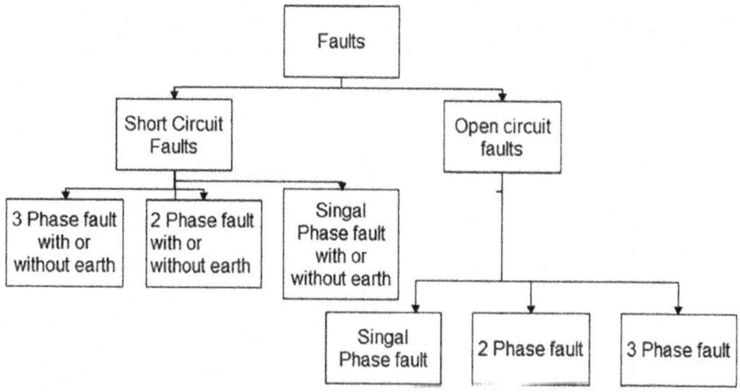

Fig. 8 various faults occurred in transmission line

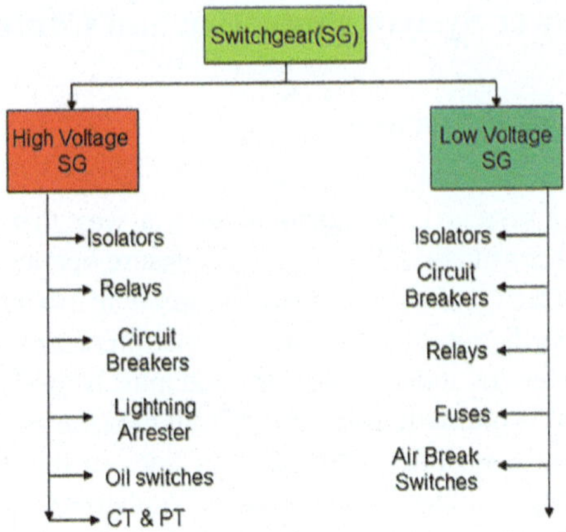

Fig. 9 Switchgears used for protection of power system

is an automatic protection system that isolates the healthy part of the system from a short circuit [12].

The design of a protection system required two types of devices, the first type of device is used to sense the value of current and voltage in the normal situation as well as abnormal situation, the second type of device is used to break the faulty section of the line from the healthy line. CT (Current transformer) and VT (Voltage transformer) are used as primary devices; they can sense the value of current and voltage. Relay is also used as a sensor that senses the value of CT current and accordingly sends information to the circuit breaker. Circuit breakers and switchgears are used as secondary devices which can break the circuit during abnormal condition. Figure 9 shows the different types of switchgears Fig. 10 shows the various relays and Fig. 11 shows various circuit breakers used for the protection of the power system [13].

The general layout of the protection system used for the electrical power system is shown in Fig. 12. CT and PT are used for the measurement of current and voltage from the power system using step down the principle of the transformer. Relay is connected with CT & PT and reads the current and voltage values and according to the value of CT and PT it takes the decision whether it is a normal operation going on or any abnormal situation arises. If an abnormal situation arises it sends a signal to the trip connector. The trip connector activated and magnetized the trip coil so that it can operate the circuit breaker. So using a trip coil circuit breaker opens the faulty parts of the system. In this way protection system protects the whole power system.

But in the classical protection and control system has many problems. Most common problem is to transfer the signal in high speed without any interference. Since classical system are using copper wire as a communication line. So in this communication system has chances of magnetic interference and loss of signals. Since copper wire is using so losses also arises [14]. For the protection of failure of system and avoid cut of power need new technology-based automatic protection

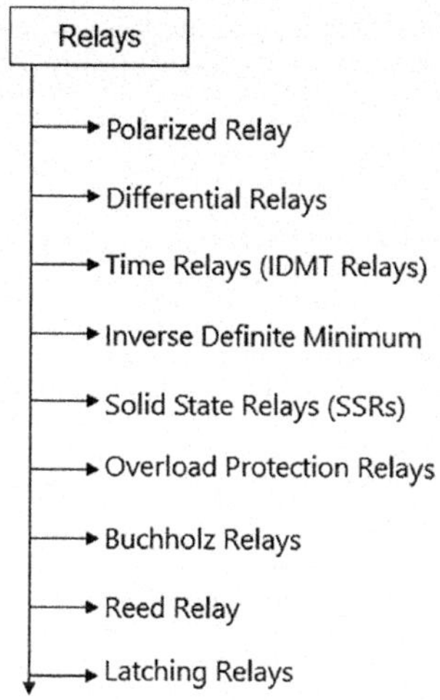

Fig. 10 Different types of relays used in power system

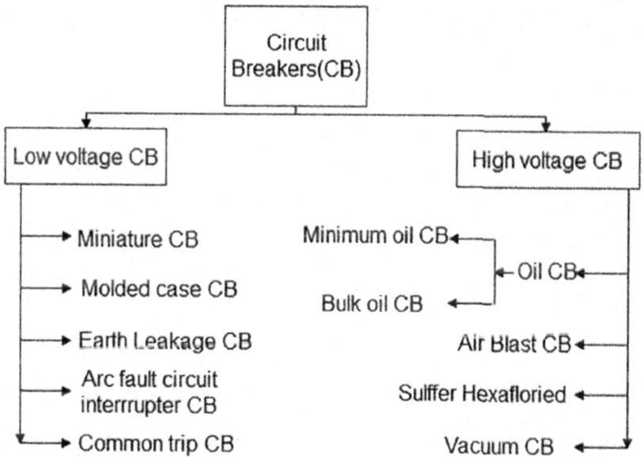

Fig. 11 Different types of CB used in power system

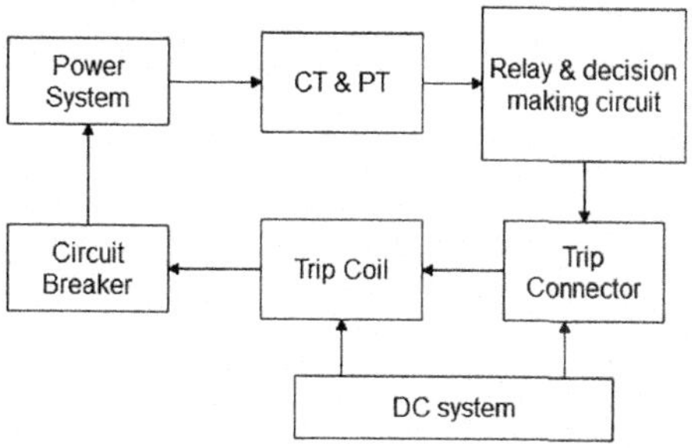

Fig. 12 Layout diagram of power system protection

system. Such an Automatic system accurately monitors, control, and protect the power system and improve the efficiency and reliability of the system.

3.2 Qualities Required of a Protection System

The protection system has the following qualities:

- Reliability
- Stability
- Selectivity or discrimination
- Sensitivity
- Adequateness
- Speed and time
- Simplicity and economy

4 Automation Power System Protection System with Fiber Optics

Automation system has an effective impact in all sectors of industries which can make them more suitable for human daily life. The automation system also increases the productivity of work, improves manufacturing quality, and decreases total costs. The use of Automation in the power system provides effective and efficient monitoring and control.

SCADA systems recently started to use in the power systems as supervisory and control units. As the signal is received from the power system switchgear accordingly SCADA system takes action and provides control and protection. In the power system for the communication of signals, a copper conductor is used. As already discussed in this chapter copper conductor communication suffers from electromagnetic interference and hence communication signal is not clear for some time. The speed of the signal is also not good. So If the signal is not clear and also not reached in time by the protective elements, how they can take proper action, and how to provide protection at the correct time?

To overcome the problem of copper conductor communication systems fiber optics can be used in place of copper conductors. As already discussed fiber optics can transmit the signal with high speed for long distances and high bandwidth without any interference. So there is no chance to lose the signal. If the protection system received the correct signal it can take corrective action in time and provide safe operation of the power system.

The optical fiber is the most important communication part between the field data interface, control units, and the SCADA system for transferring the data signal. Fiber optic communication is free from electromagnetic and radio frequency interference,

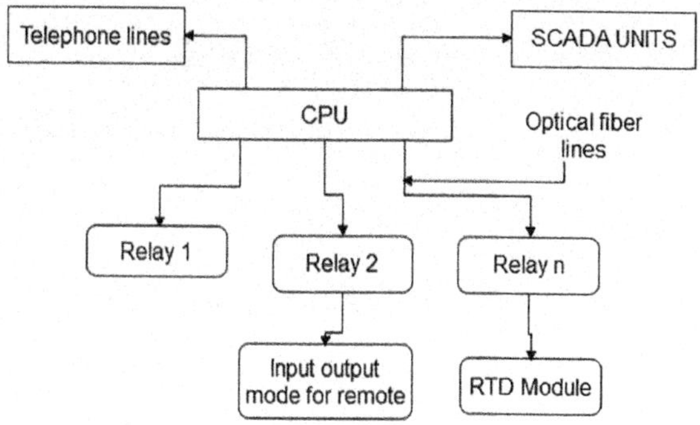

Fig. 13 SCADA system with fiber optics communication

sparking does not arise like in copper conductor, data is transited securely and very low attenuation coefficient, loss of data is very less and suitable for long communication at very low cost. The structure of SCADA with fiber optics communication system is shown in Fig. 13.

CPU is the central processing unit used to monitor all processes and send signals to different relays to trip the circuit under abnormal conditions. All relays are connected with fiber optic cables with CPU units. As with any abnormal situation observed by CPU units its send message to the SCADA system as well as different connected relays. As the SCADA system received a signal from the CPU unit it sends a signal to different protective devices by fiber optics and protective units operate and turns off the power as soon as possible. Optical technology applications in electrical power systems have in the main included the following major aspects shown in Table 4 and (Figs. 14 and 15).

Table 4 Application of optical fiber system in electrical power system measurements and communications

S. no	Application area	Measurement (%)	Communication (%)
1	Control & monitoring	40	55
2	Substations	30	70
3	Fault location	45	50
4	Unser ground cables	30	80
5	Overhead lines	10	95
6	Distribution system	45	55
7	Relay and protection	15	90

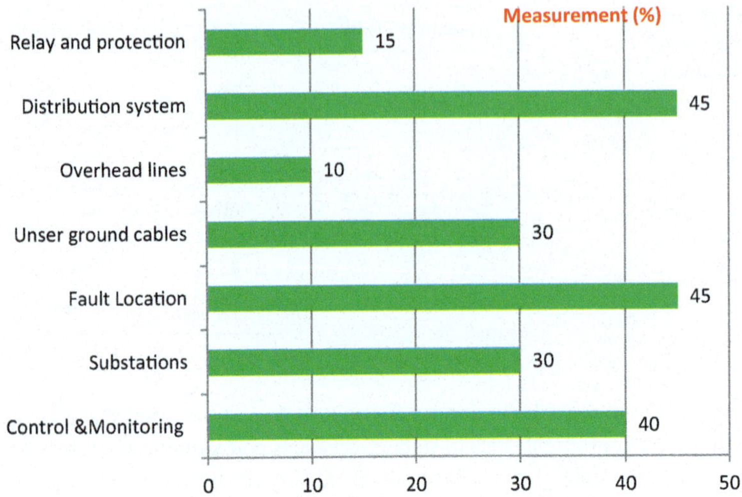

Fig. 14 Use of measurement in percentage in different power system areas

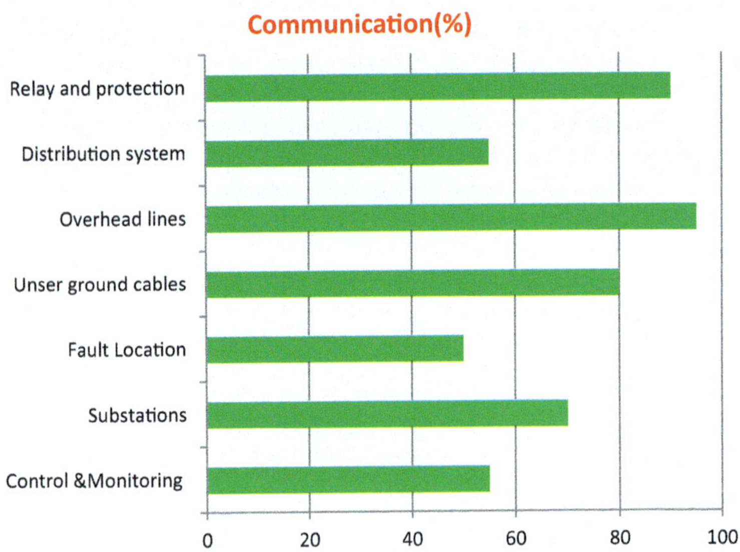

Fig. 15 Use of communication in percentage in different power system areas

5 Conclusions

The electrical power system has a very large network and continuously works for the supply of power for different consumers. A huge of capital is invested in the generation, transmission, and distribution of power. But safety is the most important phenomenon in the power system. From time to time protection and control system is updated by the electricity companies. But that is not enough as the demand and utilization are increasing continuously. The switchgear and protection system work together using communication copper conductor signals. The copper conductor communication system has many issues so the protection system also does not work accurately.

Now the time has come to update the communication system using fiber optics which has so many great advantages which make it suitable for communications. For power system protection and control, optical fiber can play a big role in providing accurate signals which high speed. So using a SCADA system with fiber optics can provide great control and protection of the power system.

References

1. M. El-Hami, K.T.V. Grattan, An overview of optical-fibre technology applications in electrical power systems. Meas. Control. **33**, 296–302 (2000)
2. D.C. Erickson, The use of fibre optics for communications, measurement and control within high voltage substations. IEEE Trans. Power Appar.Us Syst. **1**, PAS-99 (3), 1057–1063 (1980)
3. J.D. Juddleston, T.A. Phillippe, A survey of optical channels for protective relaying practices and experience. IEEE Trans. Power Deliv. **10** (2), (1995)
4. Fibre optics relay channels working group: Fibre optic channels for protective relaying. IEEE Trans. Power Deliv. **4** (1), 165–176, (1989)
5. Y.A. Ning, K.T.V. Grattan, W.N. Wang, A.W. Palmer, A systematic classification and identification of optical fibre sensors. Sensors and Actuators A **29**, 21–36 (1991)
6. Y.N.Ning, Z.P. Wang, A.W. Palmer, K.T.V. Grattan, D.A. Jackson, Recent progress in optical current sensing techniques, Rev. Sci. Instrum. **66** (5), 3097–3111 (1995)
7. H. Yu, P. Li, L. Zhang, Y. Zhu, F.A. Al-Zahrani, K. Ahmed, Application of optical fiber nanotechnology in power communication transmission. Alex. Eng. J.**59**, 5019–5030 (2020)
8. W.J. Lv, S.R. Zhai, Y.N. Qiao, Research and application of grid integration system for power energy metering and acquisition. J. China Acad. Electron. Inform. Technol. **13**, 103–108 (2018)
9. O.S. Sidelnikov, A.S. Skidin, M.P. Fedoruk, Investigation of nonlinear effects in the transmission of a qam signal in fibre optic communication lines using different carrier pulses. Quantum Electron. **47**, 1140–1143 (2017)
10. H. Helmers, C. Armbruster, von Ravenstein M, D. Derix, C. Schöner, 6-W Optical power link with integrated optical data transmission. IEEE Trans. Power Electron. **35** (8), (2020)
11. J. B. Rosolem, Quantitative and qualitative monitoring system for switchgear with full electrical isolation using fiber-optic technology. IEEE Trans. Power Delivery **30** (3), 1449–1457 (2015)
12. H. Helmers, Photovoltaic Cells with increased voltage output for optical power supply of sensor electronics," in *Proceedings of 17th InternationalConference SensorMeas.Technology*, (2015), pp. 519–524
13. H. Helmers, D. Lackner, G. Siefer, E. Oliva, F. Dimroth, A.W. Bett, Integrated power and data transceiver devices for power-by-light systems–a concept study. in Proc. 32nd Eur. Photo volt. Solar Energy Conf. Exhib. (Munich, Germany, 2016), pp. 218–222

14. J.S. N'cho 1, I. Fofana, Review of fiber optic diagnostic techniques for power transformers. Energies. **13**, 1789, 1–24 (2020)
15. M. Lindgren, M. Kharezy, Fiber optic sensors for high-voltage applications. (SP Technical Research Institute of Sweden, Bros, Sweden, 2015)

Orbital Angular Momentum of Light in Helically Twisted Hollow Core Photonic Crystal Fiber

Rik Chattopadhyay

1 Introduction

Orbital angular momentum (OAM) carrying light has found immense application in fields like particle trapping, quantum information storing, high bandwidth communication channel, and quantum communication [1]. Since the first proposal of exciting light that can carry certain OAM [2], some reports suggested methods of exciting light that carries OAM which is useful for specific application during the last three decades. However, there are not many reported results related to confinement and propagation of such OAM carrying light over a long distance [3–8]. Though some of these specially designed waveguides can transmit OAM carrying modes over a few meters [5] but the excitation of such modes in the system needs external arrangements and it requires specially designed couplers to transmit the signal from the waveguide to the peripherals.

Recently it has been shown that when a solid core photonic crystal fiber (PCF) is twisted helically along the length, polarization degeneracy of the core modes gets lifted [8–10] and the photonic crystal (PC) cladding can support Bloch modes carrying certain OAM [11]. As stated in [12, 13] the hexagonal cladding in normal PCF supports fundamental space filling mode (FSM) whose axial Poynting vector directs toward the fiber axis. When the PCF is slightly twisted the FSM is forced to follow a helical path around the axis and it picks up a transverse momentum. The discrete OAM order is then determined by the PC radius and twist rate [14]. It has also been shown that in such twisted PCF the core guided mode can resonantly couple with selected OAM carrying Bloch modes and as a result it suffers a series of dips in the output spectrum [12, 14]. It has also been shown that OAM can be preserved in the core mode if the core of the PCF is designed as a three-bladed Y-shaped structure [13].

Our idea is to avoid such special design of PCF cores so that any standard fiber optic peripherals can be used while designing a system. This is achieved when we can couple the OAM carrying Bloch modes of a twisted PC in the defect core of a

---

R. Chattopadhyay (✉)
Department of Electronics and Telecommunication Engineering, Indian Institute lf Engineering Sciencc and Technology, Shibpur, Howrah, India
e-mail: rchattopadhyay@telecom.iiests.ac.in

A. Saha et al. (eds.), *Optical to Terahertz Engineering*, Progress in Optical Science and Photonics 23, https://doi.org/10.1007/978-981-99-0228-6_8

PCF. In previous reports it has been demonstrated that in some hexagonal PCs with typical values of pitch, Λ (center to center distance of two adjacent air holes) and air hole diameter d, a conical Dirac point appears in the photonic band structures and PC modes at this Dirac point (DP) can be trapped in a centrally located air defect [15, 16]. The trapping of such PC modes in a centrally located defect, deliberately embedded in the PC, can be tuned by changing the diameter or dielectric permittivity of the defect. We try to investigate the possibility of exciting PC modes in a helically twisted hollow core PCF (HC-PCF) cladding so that some topological phases can be associated with PC modes. Then by utilizing the Dirac point resonance between PC mode and defect mode we try to couple the phase information from PC mode to the defect mode. We expect that this phase information will be preserved by the propagating defect modes in the air core of HC-PCF. Hence we can achieve the goal of exciting a guided mode carrying certain OAM in an HC-PCF.

It may be argued that the azimuthal momentum induced in the Bloch mode can be destroyed due to the transfer of momentum within the structure. Even if the Bloch modes lose their azimuthal momentum, it certainly will not hinder to couple an azimuthal momentum in the defect guided mode because the defect mode will twist to resonate with spiraling Bloch mode, but there will be no transverse structural discontinuity in the defect to destroy the azimuthal momentum.

We propose a helically twisted HC-PCF with $\Lambda = 2.21\ \mu$m and $d = 0.94\Lambda$ with a centrally located air defect of radius $r_D = 1.73\Lambda$ to excite and guide OAM modes. We have studied three different fiber with host glass as pure fused silica ($\epsilon_r = 2.1025$), GeO_2 ($\epsilon_r = 2.5281$) and SF_6 ($\epsilon_r = 3.24$). In the present case, the Bloch modes at Dirac frequency are forced to follow a helical path due to twist. Hence, the Bloch modes will rotate during propagation [14]. Now the spiraling Bloch modes are trapped in the defect when a resonance occurs between the defect mode and Bloch mode at Dirac frequency. This in turn excites a guided mode at the central hollow core. the resonance will be maintained if the defect mode started twisting at the same frequency as that of the Bloch mode. Only at such conditions the defect mode will see a consistent Dirac point during propagation. Hence the defect mode will pick up an azimuthal momentum and in turn carry OAM. The Bloch modes rotating along the twist experience a Dirac point resonant guiding mechanism. On the other hand, Bloch modes rotating against the twist cannot see the Dirac cone-like degeneracy as the translation invariance of the PC is lifted. Therefore, these modes cannot resonate with a defect mode. Since the twist induces circular birefringence between the degenerate polarizing modes [8, 17], the defect modes in helically twisted HC-PCF show transmission peaks at different wavelengths. Therefore it becomes very easy to separate the different OAM modes in the proposed fiber. The resonance band depends on the twist rate heavily. The propagation constant of the Bloch modes depends on the twist rate and thus dispersion of Dirac point becomes twist dependent. As a result the resonance between the Bloch modes and defect mode depends on the twist rate.

We use mode analysis technique in finite element method (FEM) based COMSOL multiphysics to calculate the photonic bandgap and bandgap dispersion of the PCs as considered in the present study. The dispersion and transmission loss of the defect modes at Dirac frequencies are also calculated using FEM. The transmission window for the OAM modes is also calculated using an analytical theory based on the behavior of Bloch modes in twisted crystal and the results are verified with the transmission loss calculated using FEM. The effects of twist rate, PC dimension, host glass permittivity, and geometry of the defect are studied and a universal design methodology is proposed. We show that the transmission window of such guided OAM modes can be tuned by the external twist introduced in the HC-PCF.

2 Dirac Like Bloch Modes in HC-PCF

The photonic band structure (PBS) of any PC can be calculated by solving Maxwell's wave equation over a unit cell of the PC with Bloch periodic boundary condition [18]. Implementation of Bloch periodicity discretized Maxwell equation for the electric field in the form given by [19],

$$-\sum_{\vec{G}'} k(\vec{G}-\vec{G}')(\vec{k}+\vec{G}')\times \left\{(\vec{k}+\vec{G}')\times E_{kn}(\vec{G}')\right\} = \frac{\omega_{kn}^2}{c^2}E_{kn}(\vec{G}) \tag{1}$$

where E_{kn} denotes the electric field of the nth band with wave-vector $\vec{k}$, $\vec{G}$ is the reciprocal lattice vector defined on the reciprocal lattice of the PC. Solving the eigen-value Eq. (1) one can find the allowed frequency ω_{kn} of the nth band for the wave-vector $\vec{k}$.

Equation (1) can be solved for two different conditions (i) the wave-vector has no component along the PC-normal (normal to the PC surface) (ii) the wave-vector

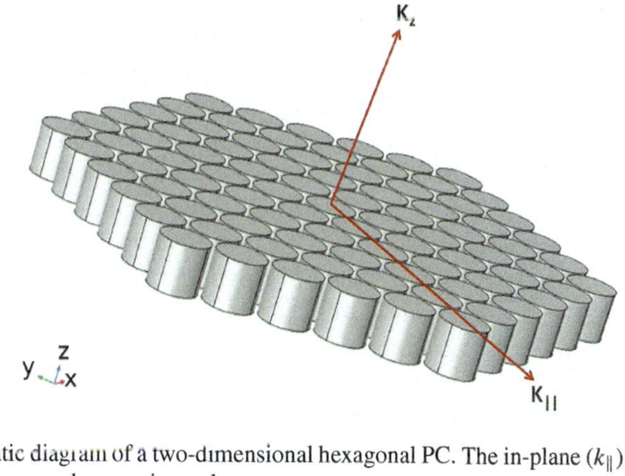

Fig. 1 Schematic diagram of a two-dimensional hexagonal PC. The in-plane ($k_{\|}$) and out-of-plane (k_z) wave-vectors are shown using red arrow

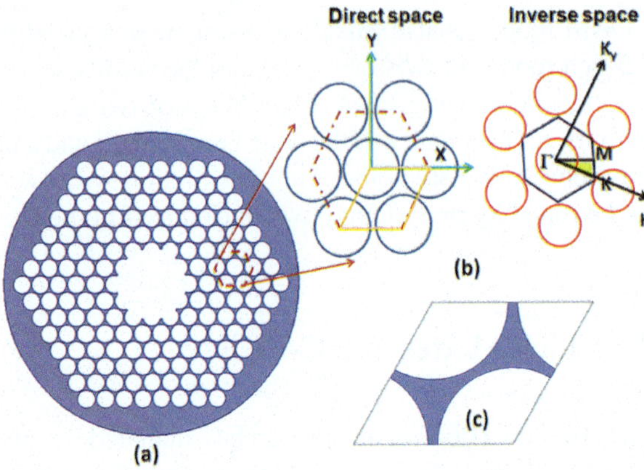

Fig. 2 **a** Cross-sectional view of an HC-PCF with a central air defect (yellow region). The blue region is the high index base glass and white holes denote the air capillary. **b** Zoom in view of the region marked by red dashed hexagon in **a**. The red dashed hexagon indicates the hexagonal lattice. The unit cell is marked by orange lines. The schematic inverse space lattice (black hexagon) is shown where the triangle filled with yellow shows the irreducible first Brillouin zone. The three highly symmetric point (Γ, M and K) are indicated. **c** The unit cell of a hexagonal PC as marked by orange lines in direct space in **b**

has a non-zero component along the PC-normal. Since we are considering a two-dimensional hexagonal PC we choose the PC plane as X-Y plane and Z-axis is the PC-normal (Fig. 1). The component of wave-vector along the PC-normal is designated as out-of-plane wave-vector k_z and the resultant component along the PC surface is designated as in-plane wave-vector $k_{\|}$. The eigen-value Eq. (1) can be solved using a plane wave expansion method. In this study, we used FEM based plane wave expansion method where the k vectors are varied over the irreducible first Brillouin zone (Fig. 2).

We varied the wave-vector k over the path $\Gamma \rightarrow M \rightarrow K \rightarrow \Gamma$ and Eq. (1) is solved on the unit cell using FEM. We have evaluated PBS for different values of k_z and the variation of the PBS with k_z is plotted to obtain a complete PBS for a given PC. The PBS calculated for the PCs considered in this study is shown in Fig. 3.

We observe that the conical nature of dispersion near the DP becomes prominent when the permittivity contrast between the base glass and air hole increases (Fig. 3a, d and g). The complete PBS plot reveals the condition of trapping of the crystal Bloch modes in the central defect. When the PC mode at DP and the defect eigen mode have the same eigen-frequency (crossing points of solid red, blue, and yellow-green lines in Fig. 3c, f and i), then these two modes can resonate and light from PC mode can be trapped in the defect. Since k_z has a non-zero value at this resonance points so the Dirac modes trapped in defect will have some component of the wave-vector along PC-normal, in this case along the fiber axis. Therefore in HC-PCF one can observe a propagating trapped Dirac mode if the defect is chosen suitably [16]. Figure 4 shows the insertion of a circular air defect in the PC. The

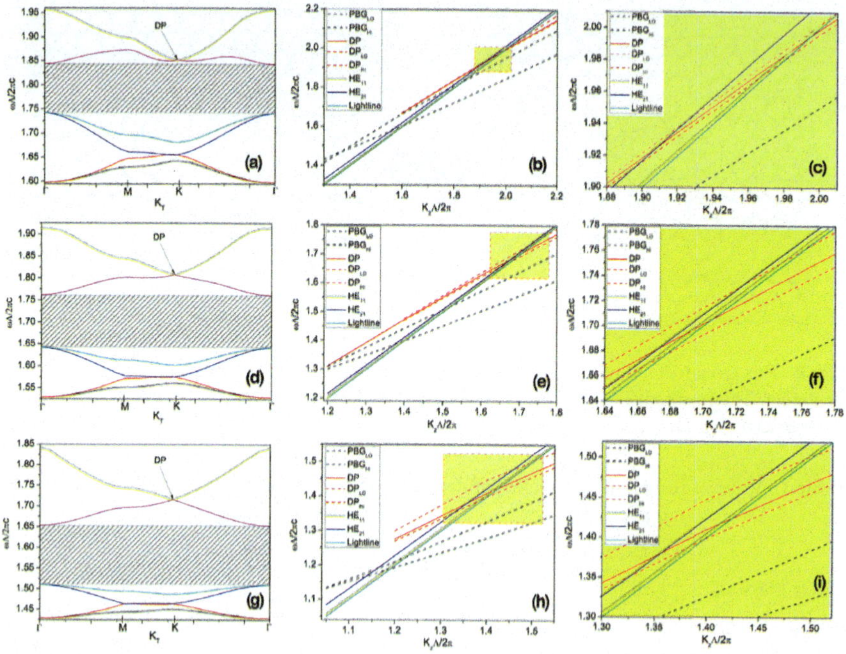

Fig. 3 **a** In-plane PBS calculated by FEM with $\frac{k_z \Lambda}{2\pi} = 1.8$ for SiO_2 glass-air hole. The Y-axis is normalized by multiplying $\Lambda/2\pi c$ where c is the velocity of light in free space and X-axis is transverse wave-vector K_T. DP indicates the Dirac point appeared in the PBS. The hashed region shows the photonic bandgap. **b** The variation of PBS with k_z is shown for the corresponding PBS. **c** Magnified plot of the region marked by yellow box in **b**. **d–f** For GeO_2 glass-air hole and **g–i** for SF_6 glass-air hole based PC, respectively. The PCs have $\Lambda = 2.21$ μm and $\frac{d}{\Lambda} = 0.94$ where d is the air hole diameter. DP_{LO} and DP_{HI} indicate the lowest and highest linear dispersion eigen-frequencies, respectively, near DP. The eigen-frequency variation with k_z of HE_{11} and HE_{21} modes of the defect is also shown for the respective cases. Lightline indicates the dispersion of light in free space

red circle shows the defect with radius $r_D = 1.73\Lambda$. To find out the defect mode we have calculated the eigen-frequency of the resultant PC geometry with defect for various values of k_z. Since we are interested in exciting OAM modes with order ± 1, we tracked both HE_{11} and HE_{21} modes. The dispersion of these two modes in all the three considered PCs is shown in Fig. 3b, e, and h. The resonance parameters between DP and defect modes for different PCs are given in Table 1.

Now we find that the normal component of wave-vector, i.e., k_z is non-zero for all resonance conditions (Table 1). This means when the PC mode resonates with a defect mode, a propagatory phase gets associated with it. Hence such eigen solutions of Maxwell's equation in the defect region will be propagatory in nature. This is relatively a new phenomenon where light can be guided in a photonic crystal fiber (PCF) beyond modified total internal reflection, as in solid core PCF and photonic bandgap guidance as in HC-PCF [14]. This beyond bandgap guidance mechanism can

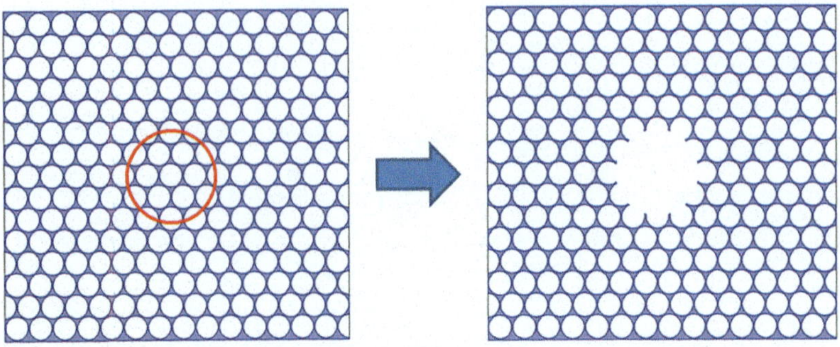

Fig. 4 Hexagonal PC with circular air defect at the center. The red circle indicates the central air defect. The resultant PC with the central defect is shown simultaneously. The defect radius is $r_D = 1.73\Lambda$

Table 1 Value of resonance parameters of defect mode with the Dirac mode of PC

	HE_{11} mode		HE_{21} mode	
	$\omega\Lambda/2\pi c$	$k_z\Lambda/2\pi$	$\omega\Lambda/2\pi c$	$k_z\Lambda/2\pi$
SiO_2 glass-air	1.9821	1.987	1.924	1.907
GeO_2 glass-air	1.696	1.693	1.683	1.673
SF_6 glass-air	1.418	1.413	1.38	1.356

be termed as Dirac resonance (DR). However, only the resonance does not guarantee the trapping of light in the defect. From Fig. 3b, e, and h we observe that Dirac-like point in PBG does not appear for any value of k_z. Now if the resonance between DP and defect mode appears on the right side of the lightline in PBS (Fig. 3c, f, and i) then trapping of light in the defect is not possible. In such cases, the trapped photon must travel faster than c, which is impossible. Hence, DR is possible only when such resonance occurs on the left side of the lightline in PBS. In the present study, we observe that DR is possible for all the considered PCs.

3 DR in Helically Twisted HC-PCF

The schematic of the twisted HC-PCF and the cross-sectional view of the actual geometry is shown in Fig. 5. The pertinent question arises in case of such twisted PC whether the Bloch mode degeneracy will hold or not? It has been established earlier that when Z-invariance for two-dimensional PC is lifted the frequency degeneracy of Bloch modes also gets lifted [8–10]. Therefore, it is expected that the Dirac-like degeneracy of the Bloch modes will be lifted in such twisted HC-PCF. But if we make a coordinate transformation from a rectangular Cartesian system (X,Y,Z) to a helical system (U,V,W) then the Z-invariance can be restored. The coordinate transformation

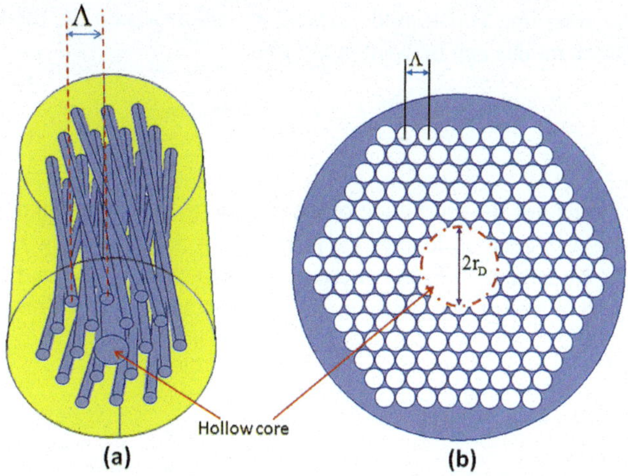

Fig. 5 **a** Schematic of the helically twisted HC-PCF. The blue cylinders represent the air capillaries. The central large capillary serves as the hollow air core. The yellow region indicates the base glass. **b** Cross-sectional view of the actual geometry considered in the study. The defect is indicated by the dashed red circle with radius $r_D = 1.73\Lambda$

relation is given by Eq. (2) as follows:

$$\begin{pmatrix} X \\ Y \\ Z \end{pmatrix} = \begin{pmatrix} cos\theta & sin\theta & 0 \\ -sin\theta & cos\theta & 0 \\ 0 & 0 & 1 \end{pmatrix} \begin{pmatrix} U \\ V \\ W \end{pmatrix} \tag{2}$$

where the angle $\Theta = \alpha z$, α is the twist rate given in unit of rad/mm and z is the length of propagation from input plane. Close inspection of Eq. (2) reveals that Maxwell's equation in such curvilinear coordinate (U,V,W) will no longer remain orthogonal, i.e., the variation of the electric field along Z-axis cannot be replaced by a simple propagation phase $e^{i\beta z}$ where β is the propagation constant.

For an untwisted crystal, the reciprocal vectors for a hexagonal PC can be written as [19],

$$\begin{pmatrix} b_1 \\ b_2 \end{pmatrix} = \frac{2\pi}{\Lambda} \begin{bmatrix} 1 & \frac{1}{\sqrt{3}} \\ 0 & \frac{2}{\sqrt{3}} \end{bmatrix} \begin{pmatrix} \hat{i} \\ \hat{j} \end{pmatrix} \tag{3}$$

where $\hat{i}$ and $\hat{j}$ are the unit vectors along X-axis and Y-axis respectively. Now if we consider a helically twisted PC with twist rate α rad/mm then the reciprocal lattice vectors can be written as using the transformation Eq. (2),

$$\begin{pmatrix} g_1 \\ g_2 \end{pmatrix} = \frac{2\pi}{\Lambda} \begin{bmatrix} \frac{cos(\frac{\pi}{6}+\alpha z)}{cos\frac{\pi}{6}} & -\frac{sin(\frac{\pi}{6}+\alpha z)}{cos\frac{\pi}{6}} \\ \frac{sin(\alpha z)}{cos(\frac{\pi}{6}+2\alpha z)} & \frac{cos(\alpha z)}{cos(\frac{\pi}{6}+2\alpha z)} \end{bmatrix} \begin{pmatrix} \hat{i} \\ \hat{j} \end{pmatrix} \tag{4}$$

Now if we consider that Bloch mode exists for such a twisted PC then the electric field of the Bloch modes can be written as [19],

$$E(\vec{r}) = \sum_{m,n} E_{m,n}(\vec{r}) e^{-i(k' + m\vec{g}_1 + n\vec{g}_2).\vec{r}} \tag{5}$$

where $E_{m,n}(\vec{r})$ denotes the amplitude of the electric field at $\vec{r}$ for (m,n) order Bloch mode. $k' = 2\pi/\lambda$ in general denotes the propagation constant along z for operating wavelength λ. We have plotted the in-plane Bloch vector $[\vec{g}_1, \vec{g}_2]$ and phase, of the mth order Bloch mode (m = n) with propagation distance Z. The plots are shown in Fig. 6.

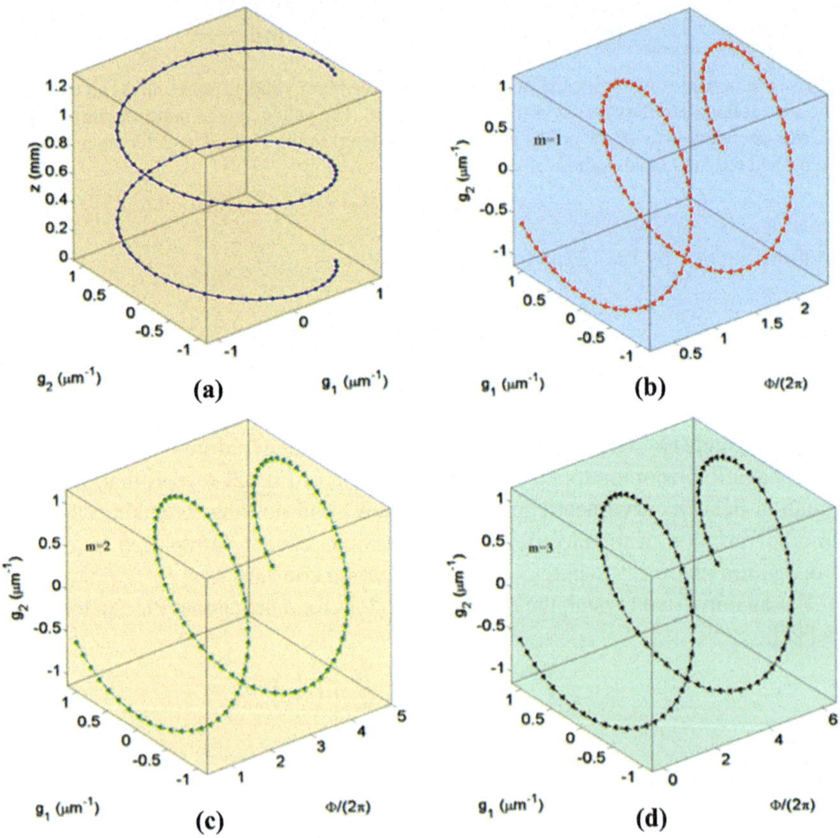

Fig. 6 **a** Tip of the Bloch vector $(\vec{g_1} + \vec{g_2})$ for each Z up to 2L where $L = 2\pi/\alpha$ is the twist pitch. Variation of the phase angle of the electric field with propagation length **b** for first higher order Bloch mode m = n = 1 **c** for second higher order Bloch mode m = n = 2 and **d** for third higher order Bloch mode m = n = 3

It is found that the Bloch vector traverses a helical path in the twisted PC. The phase associated with mth order Bloch mode varies from $0 \rightarrow 2\pi m$ over the transverse plane. Therefore, the Bloch modes in a twisted PC will carry an OAM. This indicates that though the PC appears anisotropic for the Bloch modes defined in a laboratory frame of reference, but the PC remains isotropic for the Bloch modes defined in a helical coordinate frame. Hence the DP will occur for such modes in the twisted PC. The only difference is that due to accumulation of some azimuthal phase the wave-vector of the Bloch modes will be raised by some amount, and the DR between the Bloch modes and defect modes will occur at different frequencies. The change in the effective mode index of the Bloch modes in a twisted crystal for small twist angle α is given by [14, 17],

$$n_{tw} = n_0\sqrt{1+\alpha^2 r^2} \tag{6}$$

where n_{tw} is the effective index of the Bloch mode in twisted PC and n_0 is the index in untwisted PC. r is the distance from the fiber core axis. Equation (6) indicates that the increment of Bloch mode index not only depends on the twist rate but also on the helix radius r. This intrinsically points to the fact that in twisted HC-PCF the DR will occur for a range of frequencies and this range will depend on the dimension of the PC. If we consider a PC of 7×7 cell then the allowed value of r for the Bloch mode is larger than a 5×5 cell PC. Hence DR will occur for a wider range of frequency in an HC-PCF containing seven layers of air capillaries in the cladding PC.

This phenomenon can be demonstrated by plotting the dispersion of the defect mode and the Bloch modes at DP (Fig. 7). In this case we have considered a circular centrally located defect that has a radius $r_D = 1.73\Lambda$ and the cladding PC consists of five rings with $\Lambda = 2.21\,\mu$m. Hence the maximum value of helix radius r for the Bloch mode is $r_{max} < 7\Lambda \sim 15\,\mu$m and the minimum helix radius will be $r_{min} > 2\Lambda \sim 5\,\mu$m.

Figure 7 reveals that the DR occurs at a particular frequency for untwisted case. As soon as twist is introduced in the system, the effective mode index of the Bloch modes at DP takes a range of values depending on the value of r ranging from 5 μm to 15 μm (Eq. 6). As a result the defect mode can resonate over a wide range of frequencies. We also observe that when the dielectric permittivity contrast is high in PC, the dispersion of Dirac modes of the PC gets flattened and the DR region for different defect modes becomes well separated.

Analytical solution of $E(\vec{r})$ is not possible in the twisted frame because we defined the electric field in a non-orthogonal curvilinear coordinate facilitating FEM to solve for the electric field. We consider the unit cell as shown in Fig. 2(c) where the DP appears at $k : (4\pi/3\Lambda, 0)$ point of the irreducible Brillouin zone for all considered PCs. Therefore, during FEM simulation we set the Flouquet periodicity at the unit cell boundaries with periodic vector value $\vec{k} = 4\pi/3\Lambda\hat{i} + 0\hat{j}$. The dispersion of the defect mode is calculated using the same method. The only difference is instead of taking a single unit cell we consider a 14×14 supercell with a defect inserted at the center as shown in Fig. 4. The electric field distributions at 1115 nm (DR point for

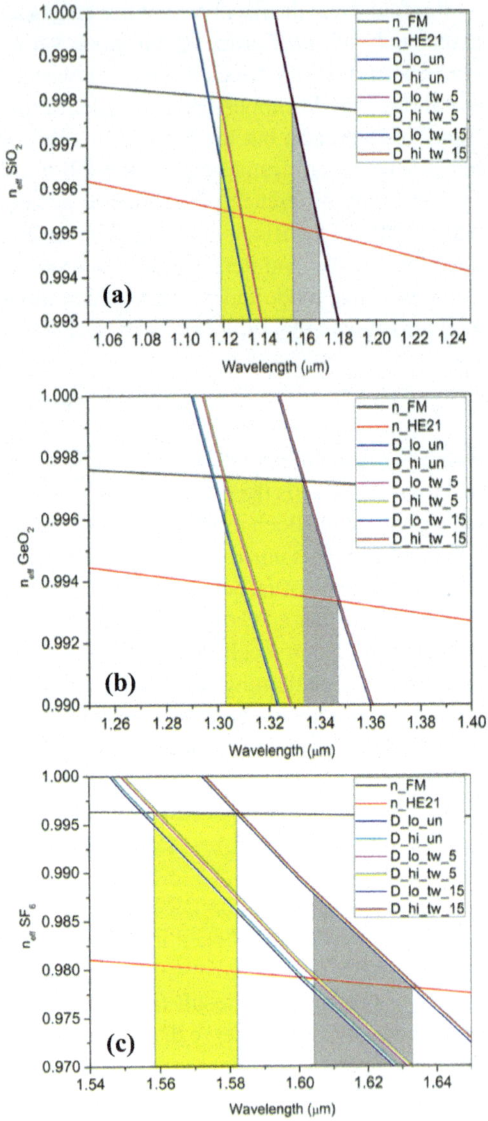

Fig. 7 Dispersion of the fundamental (n_FM) and HE_{21} (n_HE21) defect mode and PC modes at DP with and without twist calculated using FEM [12, 16] and Eq. 6. **a** SiO_2 glass-air PC **b** GeO_2 glass-air PC and **c** SF_6 glass-air PC. D_lo_un and D_hi_un denote the 5th and 6th PC band at Dirac point for untwisted case, respectively. D_lo_tw_5 and D_hi_tw_5 denote the PC bands at DP in twisted PC with helix radius $r = 5\,\mu m$. D_lo_tw_15 and D_hi_tw_15 denote the PC bands at DP in twisted PC with helix radius $r = 15\,\mu m$. The yellow region shows the DR region with fundamental defect mode and gray region denotes the DR region with HE_{21} defect mode

fundamental mode) for 5th and 6th PC bands and the fundamental defect mode for SiO_2 glass-air PC are shown is Fig. 8. After calculating the effective indices of the two PC modes at Dirac point and the defect modes for the untwisted case we have estimated n_{tw} of the corresponding PC modes using Eq. (6). These results indicate that if we introduce a defect at the center of the PC in the form of a hollow core in a twisted PCF then we may create DR over a wavelength region as compared to untwisted HC-PCF. Since the modes in twisted PC would carry certain OAM then it is expected that upon DR the defect mode will also carry certain OAM. Therefore, it is possible to trap the light from the PC modes at Dirac point in a twisted HC-PCF. It is also expected that since DR occurs for non-zero value of k_z the trapped modes would be propagating in nature.

4 OAM Modes in Twisted HC-PCF

The nature of the trapped light in the hollow core of a twisted HC-PCF is now investigated using FEM. We have already discussed the form of Maxwell's wave equation in a twisted system in Sect. 1. The non-orthogonality of the coordinate frame prevents to use the method of separation of variables to calculate the components of the electric field using a set of orthogonal equations. Hence no analytical solution to Maxwell's wave equation in the helical coordinate system is possible. But it can be shown that if we change the material properties from an isotropic to the anisotropic system using the transformation relation [7, 12, 20],

$$\epsilon' = \epsilon T^{-1} \text{ and } \mu' = \mu T^{-1} \tag{7}$$

where,

$$T^{-1} = \begin{pmatrix} 1+\alpha^2 v^2 & -\alpha^2 uv & -\alpha v \\ -\alpha^2 uv & 1+\alpha^2 u^2 & \alpha u \\ -\alpha v & \alpha u & 1 \end{pmatrix} \tag{8}$$

the wave equation can be reduced to a two-dimensional problem and now it can be solved using FEM. We have used a perfectly matched layer (PML) to calculate the confinement loss of the core guided modes. The geometrical cross section of the structure for FEM simulation is shown in Fig. 9. The Poynting vector of the fundamental core guided mode at DR for the three PC systems is shown in Fig. 9. OAM modes in such twisted PCF can be excited using the DR. We have already explained that in a twisted HC-PCF the Bloch mode follows a helical path and therefore carries certain orders of OAM. Now if we design the central defect in such a way that the HE_{mn} for $m \geq 2$ in the core shows a DR with any of the OAM carrying Bloch modes then the core mode will carry an OAM by virtue of the topology. We have simulated modal propagation for the twisted HC-PCF using FEM analysis where the permittivity tensor of the air capillaries was transformed using Eqs. (7) and (8) to capture their relative displacement in the glass. Figure 7 depicts that the HE_{21} mode in the defect shows DR from $1.14-1.17\,\mu$m in SiO_2 glass-air PC, $1.34-1.36\,\mu$m in GeO_2 glass-air PC and for SF_6 glass-air PC the DR region is in between $1.61-1.62\,\mu$m. The operating wavelengths are chosen from these resonance wavelength regions as shown in Fig. 7. The transverse field distribution plots for the defect modes with the lowest confinement loss in these regions are shown in Fig. 10.

Figure 10 shows that the DR between PC modes and different defect modes is possible in the twisted HC-PCF. The phase distribution clearly reveals a phase

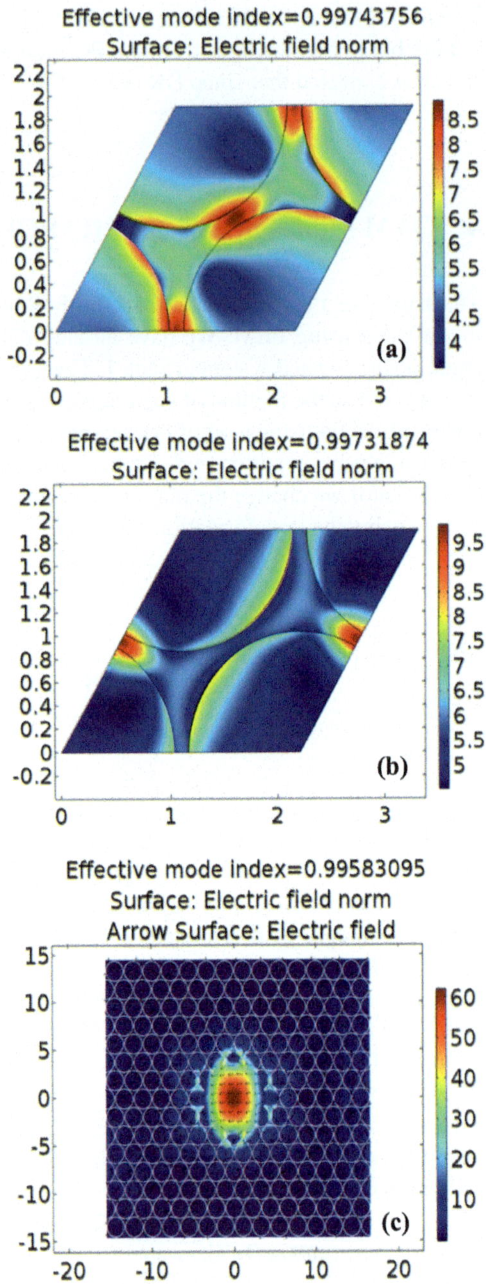

Fig. 8 Electric field distribution as generated by FEM analysis for **a** lower Dirac band **b** higher Dirac band and **c** fundamental defect mode at K point. The X-Y axes are shown in m scale

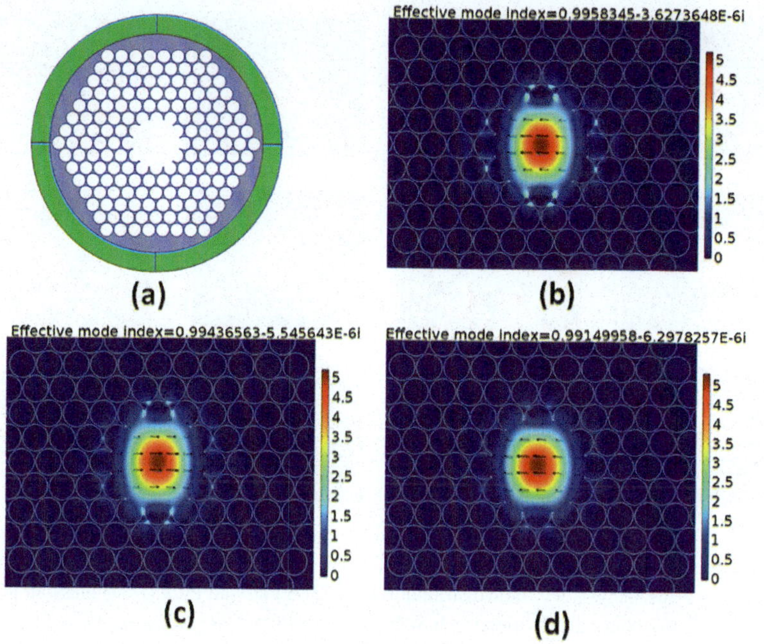

Fig. 9 **a** Cross-sectional view of the fiber geometry used for FEM analysis. The blue region is the high index glass, the white holes are the air capillaries and the green region is the PML layer. **b** Fundamental mode at 1115 nm for SiO_2 glass-air PCF **c** Fundamental mode at 1305 nm for GeO_2 glass-air PCF **d** Fundamental mode at 1575 nm for SF_6 glass-air PCF. The wavelengths are chosen at DR wavelength from Fig. 7 for untwisted cases

singularity exists at the center of the V_{21}^{+} modes of the defect. We have ensured that the tracked defect modes indeed carry an OAM by plotting the x and y components of the transverse field (Fig. 11) as shown in Fig. 10. Any OAM carrying mode in an optical fiber can be written as [6],

$$V_{21}^{\pm} = HE_{21}^{even} \pm jHE_{21}^{odd} \tag{9}$$

where $j = \sqrt{-1}$. This means, unlike a cylindrical vector mode which has linearly polarized orthogonal components, the components of any OAM mode in an optical fiber will have two HE modes superposed with a phase difference of $\pi/2$. To testify this we have first simulated the modes in the proposed HC-PCF with the help of FDTD solver (Lumerical Mode) in helical frame. We chose the operating wavelength as shown in Fig. 10. Then we decomposed the modes in its 'x'and 'y'components (Fig. 11). Since the OAM modes will be generated due to superposition of HE_{lm} modes so the 'x'and 'y'components of the electric field generated by the numerical engine will denote the constituent HE modes. Figure 11a shows the x and y components of the intensity distribution of the first OAM mode, i.e., OAM_0^{+} higher order defect mode. Since this mode can be written as $OAM_0^{+} = HE_{11}^{even} + jHE_{11}^{odd}$ so the

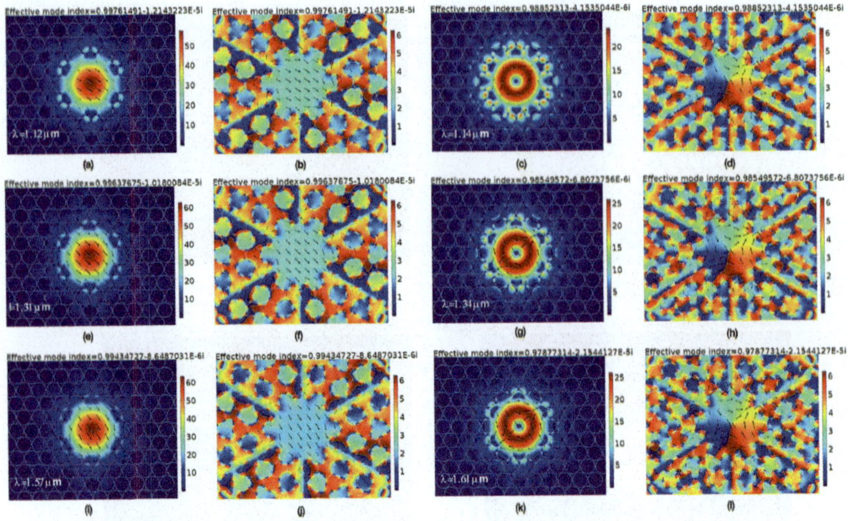

Fig. 10 Electric field distribution of V_{11}^{+} and V_{21}^{+} mode along with the transverse phase distribution obtained by FEM analysis. V_{11}^{+} field distribution for **a** SiO_2 glass-air PC **e** GeO_2 glass-air PC **i** SF_6 glass-air PC. V_{11}^{+} phase distribution for **b** SiO_2 glass-air PC **f** GeO_2 glass-air PC **j** SF_6 glass-air PC. V_{21}^{+} field distribution for **c** SiO_2 glass-air PC **g** GeO_2 glass-air PC **k** SF_6 glass-air PC. V_{21}^{+} phase distribution for **d** SiO_2 glass-air PC **h** GeO_2 glass-air PC **l** SF_6 glass-air PC. The operating wavelengths are shown in the insets. The normalized eigen-frequency can be calculated from the operating wavelength using the relation $\omega\Lambda/2\pi c = \Lambda/\lambda$ which is 1.973 for **a**, 1.939 for **c**, 1.687 for **e**, 1.649 for **g**, 1.408 for **i** and 1.373 for **k**. The twist rate is set to $\alpha = -9.6$ rad/mm, i.e., twist in a clockwise direction for all the cases. The black arrows indicate the polarization of the modes

real part of 'x'component and imaginary part of 'y'component of the electric field will exist as shown in the Fig. 11a. The Phase plot of the 'x'and 'y'component also supports this. The phase associated with 'x'component of the field is zero and that of the 'y'component is $\pi/2$ as shown in Fig. 11a. Similar analysis is done for the first higher order OAM mode as shown in Fig. 10k. The first higher order OAM mode can be decomposed as $OAM_1^{+}1 = HE_{21}^{even} + jHE_{21}^{odd}$. So the 'x'and 'y'component of the electric field will be a HE_{21} mode as shown in Fig. 11b. To ensure that the components are really a cylindrical vector mode we further plotted the rel and imaginary part of each component which reveals that both the ''x'and 'y'component of the field is formed by the superposition of two LP_{11} modes with $\pi/2$ phase difference as presented in Fig. 11b. The phase plot further confirms that both the 'x'or 'y'component have a spiraling phase, which is the signature of $OAM_1^{\pm}$ modes.

The phase distribution in PC cladding (Fig. 10b), indicates that the phase varies from 0 to 2π with 6 fold periodicity. Since we have considered a hexagonal PC, which belongs to a C_{6v} point group, the Bloch modes in PC should carry an OAM with 6 fold symmetry [14, 19]. Therefore, it is confirmed that the twisted PCF will support OAM carrying Bloch modes in the PC cladding.

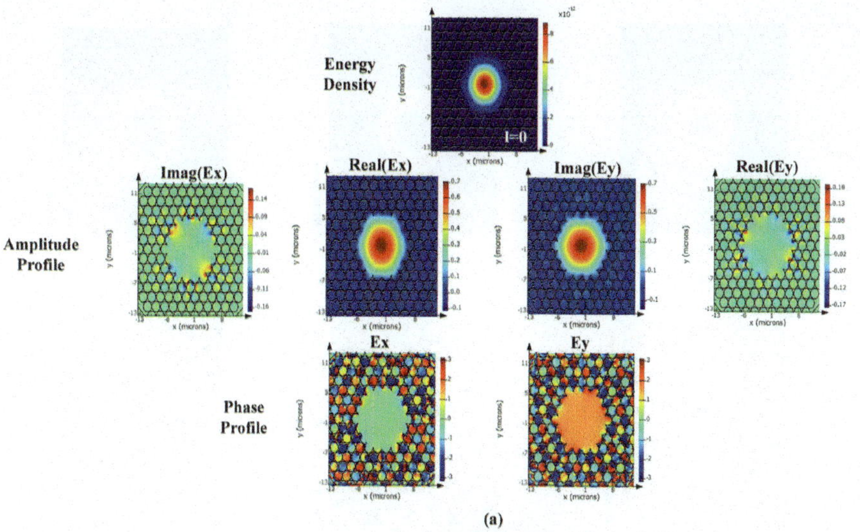

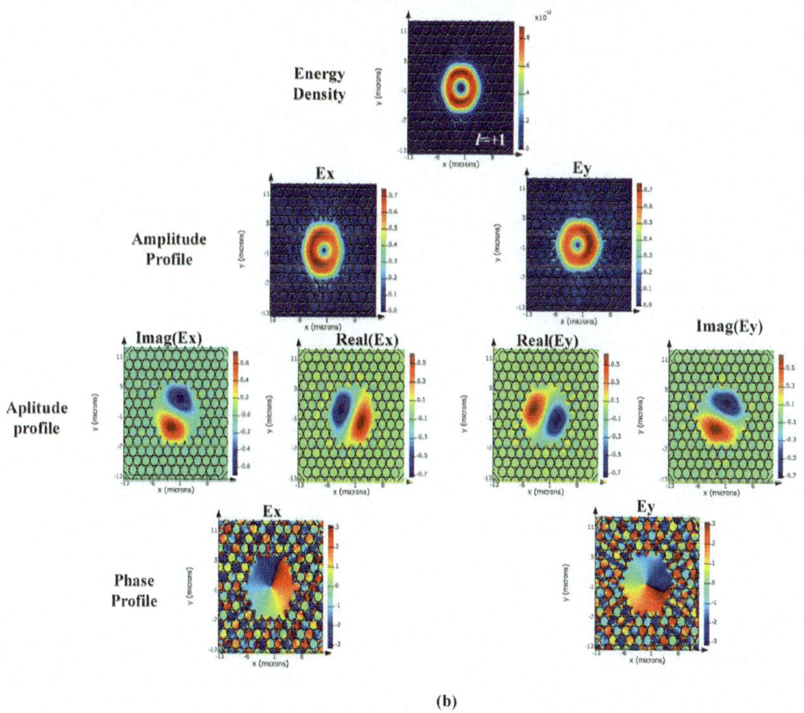

Fig. 11 Transverse electric field energy distribution for OAM_0^+ and OAM_1^+ mode in twisted HC-PCF with SF_6 host glass at 1.57 μm and 1.6 μm respectively. Transverse energy distribution, x and y component of the electric field distribution, and phase of the 'x' and 'y' components. **a** for OAM_0^+ mode **b** for OAM_1^+ mode

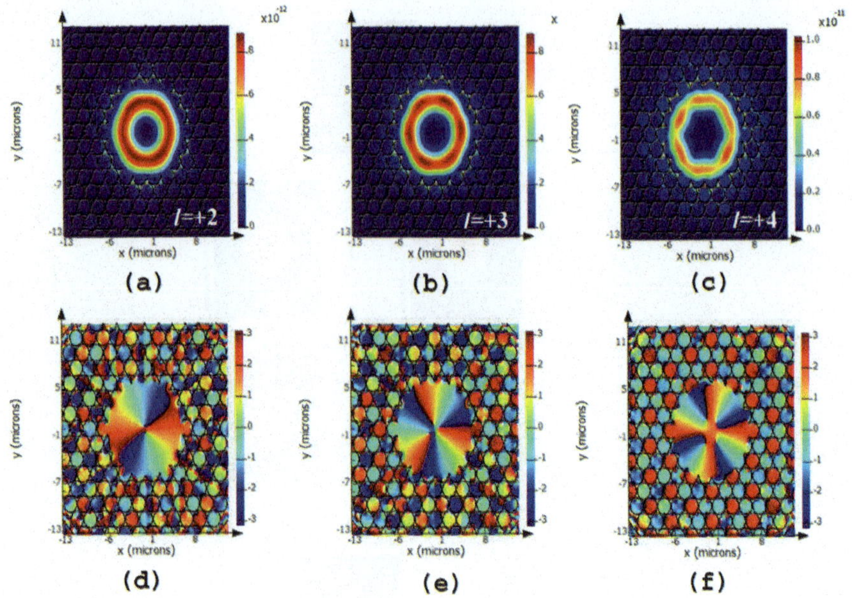

Fig. 12 **a–c** Transverse energy distribution of higher order OAM modes at 1.61 μm in the SF_6 glass based proposed twisted HC-PCF. **d–f** Respective phase profiles

It is also to be noted that other higher order OAM modes are likely to be generated inside the defect. So we tried to identify some other higher order OAM modes in the proposed twisted HC-PCF. Some of these modes are shown in Fig. 12. It is to be noted that these higher order OAM modes will have transmission peaks at different wavelengths as evident from the resonance condition (Fig. 7). OAM modes with negative OAM order will also get excited but they suffer a high loss as the Bloch modes for such negative order OAM modes propagates against the twist.

The confinement loss is calculated using the relation [21],

$$\gamma = \frac{2\pi}{\lambda}\text{Im}\{n\} \tag{10}$$

where Im$\{n\}$ denotes the imaginary part of the effective refractive index. Using Eq. (10) we have calculated the transmittance of the fundamental and HE_{21} mode in untwisted HC-PCF and OAM modes (V_{11}^{+} and V_{21}^{+}) with order $l = 0, +1$ modes in twisted HC-PCF. The transmittance plots are shown in Fig. 13.

Figure 13 shows that when the PCF is twisted the transmission windows of the V_{11}^{+} and V_{21}^{+} mode spread out. This can be correlated with the resonance plots as shown in Fig. 7. We observe that for twisted HC-PCF the Bloch modes participating in Dirac degeneracy show resonance with different defect modes for a band of wavelengths. Therefore, we expect that in twisted HC-PCF the maximum transmission will occur over this wavelength range (Fig. 13). The FWHM of transmission window increases

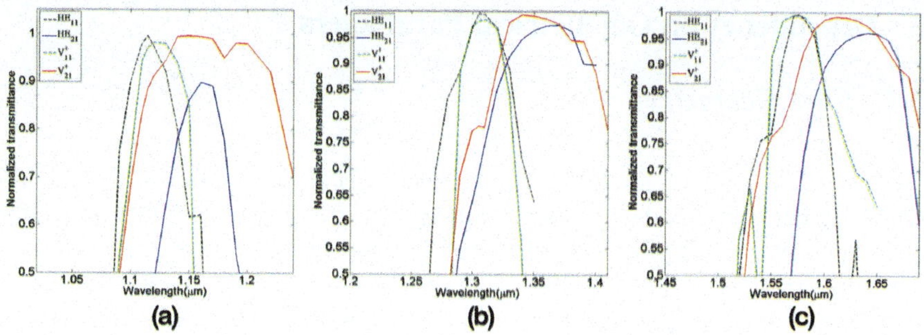

Fig. 13 Transmittance plot of HE_{11} and HE_{21} defect modes for untwisted and V_{11}^{+} and V_{21}^{+} for twisted HC-PCF calculated using FEM **a** SiO_2 glass-air PC **b** GeO_2 glass-air PC **c** SF_6 glass-air PC

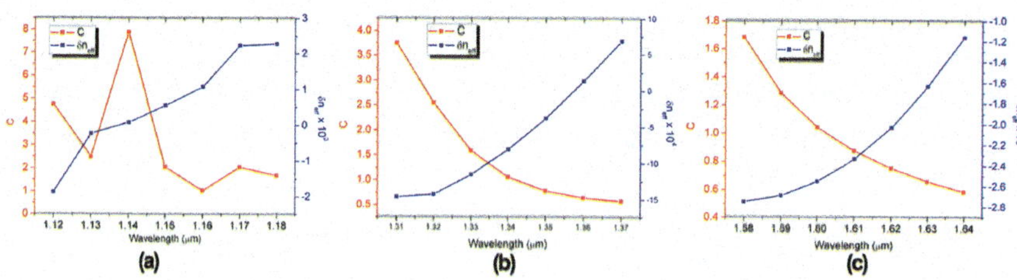

Fig. 14 Variation of C and δn_{eff} with wavelength for **a** SiO_2 glass-air PC **b** GeO_2 glass-air PC **c** SF_6 glass-air PC

for the twisted case for all considered HC-PCFs. The purity of the OAM mode can be defined as the ratio of total power carried by two OAM modes with opposite order and their index difference with parasitic TM_{01} or TE_{01} modes, δn_{eff} given by [22–24],

$$C = \frac{P_o^{RCP}}{P_o^{LCP}} \text{ and } \delta n_{eff} = n_{eff}^{TM_{01}} - n_{eff}^{RCP} \tag{11}$$

where P_o signifies the output power of the right circular polarized (RCP) and left circular polarized (LCP) OAM modes and n_{eff} signifies effective mode index of the TM_{01} and right circular polarized (RCP) mode respectively. The variations of OAM purity with wavelength for all three PCFs are shown in Fig. 14. We observe (Fig. 14) that the distribution of power between modes with OAM order +1 and −1 becomes almost equal when the base glass index increases. On the other hand, the index difference δn_{eff} between the V_{21}^{+} mode and the parasaitic TM_{01}^{+} mode is greater than 2×10^{-3} for high index glass based PC. This value is larger than conventional polarization maintaining fibers. Hence we may conclude that OAM mode excitation will not be hampered by inter-modal coupling in twisted HC-PCF. Hence these fibers are suitable for OAM excitation and guiding.

5 Dependency on Geometrical Parameters

5.1 Core Geometry

We have considered a centrally located circular defect placed in the PC to study the OAM excitation and guidance. Such a simple geometry is good for initial estimation and understanding the physical principles. In practice fabrication of such inverse floral shape (Fig. 9a) is bit difficult to achieve. Therefore we have extended our study for a more realistic core geometry which is found in regular HC-PCF, fabricated by stack and draw method. The cross section of the geometry along with a mode field distribution for one of the considered twisted HC-PCF is shown in Fig. 15.

It is evident (Fig. 15) that the nature of the trapped modes does not change with the change in shape of the hollow core. But the presence of a thin glass layer around the core increases the transmission loss of the trapped modes. As a result the DR wavelength shows a red shift. This is a very trivial phenomenon. Thin glass around the air core acts as Fabrey-Perot cavity. Hence the intensity of the surface modes gets enhanced and thus the transmission loss of the trapped mode increases. The

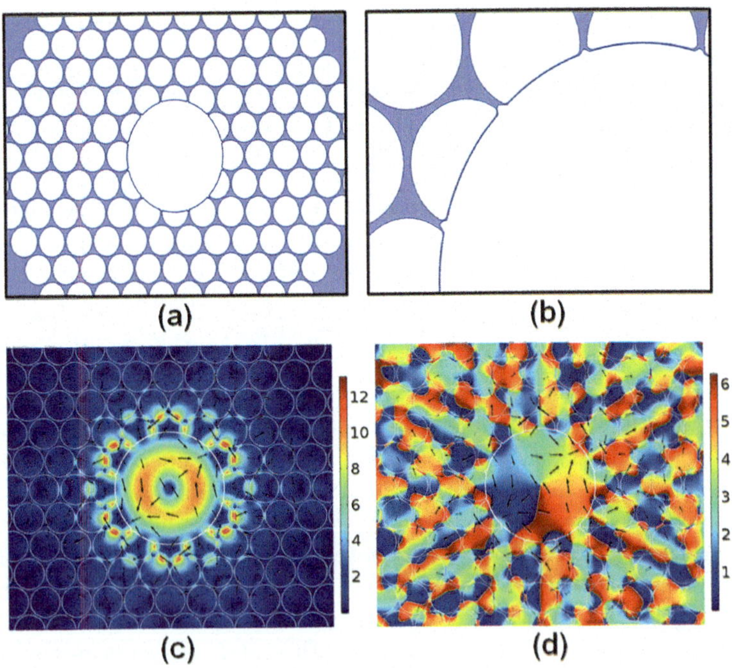

Fig. 15 **a** Cross section of the twisted HC-PCF with circular core. **b** Magnified view of the core boundary. **c** Mode field distribution of the OAM carrying HE_{21} mode for SiO_2 glass-air PC at 1115 nm. **d** Phase distribution of the mode

Table 2 Propagation loss and DR wavelength for floral and circular core HC-PCF

	Floral core		Circular core	
	DR wavelength	Loss (dB/cm) (μm)	DR wavelength	Loss (dB/cm) (μm)
SiO_2 glass-air	1.15	0.984	1.16	2.83
GeO_2 glass-air	1.34	1.38	1.35	4.35
SF_6 glass-air	1.61	3.65	1.62	12.45

propagation loss of the OAM carrying mode at DR calculated by FEM for the three chosen HC-PCFs is given in Table 2.

Therefore a special attention must be given in designing the core of the twisted HC-PCF. If we reduce the amount of glass around the core, we would achieve maximum propagation of such OAM carrying modes in a twisted HC-PCF.

5.2 *PC Size*

PC size, i.e., the number of air capillaries plays an important role in altering the confinement loss of the hollow core guided mode in conventional photonic bandgap HC-PCF. Therefore it is expected that if we increase the number of capillary layers in the PC cladding then the OAM modes will show lesser transmission loss. In this study we consider 5 layers of air capillaries in the cladding PC. We have increased the number of capillary layers from 5 to 7 in the PC. The transmission loss of the OAM modes in the twisted HC-PCF with 7 capillary layers at DR point is given in Table 3.

We observe that when the layer number is increased the confinement loss of the OAM carrying modes decreases and shows a little red shift. This can be explained based on results plotted in Fig. 7. When the ring number is increased, the upper wavelength limit of the OAM pass band in twisted fiber will show a red shift. Therefore the maximum DR point will be shifted owing to large values of PC radius. But we must keep in mind about the structural integrity while increasing the number of air rings in PC. Twist may generate tensile stress that would change the shape

Table 3 Propagation loss and DR wavelength for 5 and 7 layers of air capillaries in PC with floral core

	5 layers		7 layers	
	DR wavelength (μm)	Loss (dB/cm)	DR wavelength (μm)	Loss (dB/cm)
SiO_2 glass-air	1.15	0.984	1.17	0.207
GeO_2 glass-air	1.34	1.38	1.35	0.236
SF_6 glass-air	1.61	3.65	1.62	0.922

of the air capillaries from circular to elliptical and will destroy the isotropic nature of the PC. Since we are considering a PC with almost 97% air filling fraction, so incorporation of 8 or more rings might lead to twist induced crack in the HC-PCF during fabrication. So an optimization is needed during fabrication.

6 Conclusion

We showed that OAM carrying modes with acceptable purity could be excited and transmitted through a suitably designed twisted HC-PCF. Our study exemplified that a new guidance mechanism can take place in a twisted HC-PCF where OAM carrying Bloch modes first get excited in the PC cladding. Satisfying the Dirac resonance condition such Bloch modes can be trapped in a centrally located air core. The process convincingly shows that Dirac cone-like degeneracy of Bloch modes can be preserved in a twisted PC under rotating frame of reference. This in turn induces a topological phase in the PC modes. As a result this OAM information can be extracted and transmitted via a guiding channel. We have analyzed the effect of different fiber parameters on the transmission of OAM carrying modes of particular order. We provided a guideline of fabricating such HC-PCFs using standard fiber fabrication techniques. Our design is universal and compatible with standard fiber optic peripherals. This type of optical fiber will provide a tool for generating and transmitting OAM carrying lights, particle trapping, and quantum communication.

References

1. J.P. Torres, L. Torner, *Twisted Photons: Applications of Light with Orbital Angular Momentum*, 2nd edn. (Wiley-VCH, Germany, 2011)
2. L. Allen, M.W. Beijersbergen, R.J.C. Spreeuw, J.P. Woerdman, Orbital angular momentum of light and the transformation of Laguerre-Gaussian laser modes. Phys. Rev. A **45**, 8185 (1992)
3. A. Tandjè, J. Yammine, M. Dossou, G. Bouwmans, K. Baudelle, A. Vianou, E.R. Andresen, L. Bigot, Ring-core photonic crystal fiber for propagation of OAM modes. Opt. Lett. **44**, 1611–1614 (2019)
4. C. Brunet, P. Vaity, Y. Messaddeq, S. LaRochelle, L. Rusch, Design, fabrication and validation of an OAM fiber supporting 36 states. Optics Express **22**, 26117 (2014)
5. P. Gregg, P. Kristensen, S. Ramachandran, Conservation of orbital angular momentum in air-core optical fibers: erratum. Optica. **4**, 1115 (2017)
6. N. Bozinovic, S. Golowich, P. Kristensen, S. Ramachandran, Control of orbital angular momentum of light with optical fibers. Opt. Lett. **37**, 2451 (2012)
7. J. Ye, Y. Li, Y. Han, D. Deng, Z. Guo, J. Gao, Q. Sun, Y. Liu, S. Qu, Excitation and separation of vortex modes in twisted air-core fiber. Opt. Express **24**, 8310 (2016)
8. T. Weiss, G. Wong, F. Biancalana, S. Barnett, X. Xi, P. Russell, Topological Zeeman effect and circular birefringence in twisted photonic crystal fibers. J. Opt. Soc. Am. B. **30**, 2921 (2013)
9. L. Chen, W. Zhang, T. Yan, L. Wang, J. Sieg, B. Wang, Q. Zhou, L. Zhang, Photonic crystal fiber polarization rotator based on the topological Zeeman effect. Opt. Lett. **40**, 3448–3451 (2015)

10. G. Wong, X. Xi, M. Frosz, P. Russell, Enhanced optical activity and circular dichroism in twisted photonic crystal fiber. Opt. Lett. **40**, 4639–4642 (2015)
11. S. Longhi, Bloch dynamics of light waves in helical optical waveguide arrays. Phys. Rev. B - Condens. Matter Mater. Phys. **76** (2007)
12. G. Wong, M. Kang, H. Lee, F. Biancalana, C. Conti, T. Weiss, P. Russell, Excitation of orbital angular momentum resonances in helically twisted photonic crystal fiber. Science. **337**, 446–449 (2012)
13. X. Xi, G. Wong, M. Frosz, F. Babic, G. Ahmed, X. Jiang, T. Euser, P. Russell, Orbital-angular-momentum-preserving helical Bloch modes in twisted photonic crystal fiber. Optica. **1**, 165 (2014)
14. P. Russell, R. Beravat, G. Wong, Helically twisted photonic crystal fibres. Philos. Trans. R. Soc. A **375**, 1–18 (2017)
15. K. Xie, W. Zhang, A. Boardman, H. Jiang, Z. Hu, Y. Liu, M. Xie, Q. Mao, L. Hu, Q. Li, T. Yang, F. Wen, E. Wang, Fiber guiding at the dirac frequency beyond photonic bandgaps. Light: Sci. Appl. **4**, e304 (2015)
16. T. Biswas, R. Chattopadhyay, S. Bhadra, Dirac-mode guidance in silica-based hollow-core photonic crystal fiber with high-index dielectric rings. Phys. Status Solidi B **9**, 1–9 (2016)
17. R. Beravat, G. Wong, M. Frosz, X. Xi, P. Russell, Twist-induced guidance in coreless photonic crystal fiber: A helical channel for light. Sci. Adv. **2**, e1601421–e1601421 (2016)
18. A. Maradudin, A. McGurn, Out of plane propagation of electromagnetic waves in a two-dimensional periodic dielectric medium. J. Modern Opt. **41**, 275–284 (1994)
19. K. Sakoda, Optical Properties of Photonic Crystals, 2nd edn. Springer (2004)
20. F. Zolla, G. Renversez, A. Nicolet, B. Kuhlmey, S. Guenneau, D. Felbacq, *The Foundations of Photonic Crystal Fibers* (Imperial College Press, 2005)
21. S. Jennings, R. Pinnick, J. Gillespie, Relation between absorption coefficient and imaginary index of atmospheric aerosol constituents. Appl. Opt. **18**, 1368–1371 (1979)
22. L. Wang, A. Corsi, L. Rusch, S. Larochelle, *Investigation of orbital angular momentum mode purity in air-core optical fibers, in 2016 IEEE Photonics Society Summer Topical Meeting Series, SUM 2016* (Newport Beach, IEEE, 2016), pp.203–204
23. N. Bozinovic, Y. Yue, Y. Ren, M. Tur, P. Kristensen, H. Huang, A. Willner, S. Ramachandran, Multiplexing in fibers. Science **340**, 1545–1548 (2013)
24. L. Haisu, G. Ren, Y. Lian, B. Zhu, M. Tang, Y. Zhao, S. Jian, Broadband orbital angular momentum transmission using hollow-core photonic bandgap fiber. Opt. Lett. **41** (2016)

Review of Various Codes and Transmitter–Receiver Architecture Used in Optical Code Multiple Access System

Somali Sikder and Shila Ghosh

1 Introduction

In an optical network the one dimensional (1-D) unipolar optical orthogonal codes or pseudo orthogonal codes are used in coherent OCDMA systems to achieve spread spectrum modulation for user information bit. For proper detection of user information same set of orthogonal codes are used at the receiver end. The 1-D optical unipolar code is a string of binary bits having code length 'n' and code weight 'w' with code weight much smaller than code length i.e. w << n. To be a perfect orthogonal code from a particular code set cross-correlation among two different pair of codes should be minimum as described in the latter part of this chapter.

Traditionally these coding systems are presented as binary number sequences. This chapter also explains thc conventional techniques to derive the auto-correlation and cross-correlation constraints. It is observed that for a particular code length and weight auto-correlation constraints lies between '1' to 'w-1' for all different possible sets. The set containing different code patterns can be realized with well-defined code structures.

S. Sikder (✉)
Department of Electronics and Communication Engineering, B. P. Poddar Institute of Management & Technology, Kolkata, West Bengal 7000052, India
e-mail: somali.sikder@bppimt.ac.in

S. Ghosh
Department of Electronics and Communication Engineering, St. Thomas' College of Engineering and Technology, Kolkata, West Bengal 700023, India

A. Saha et al. (eds.), *Optical to Terahertz Engineering*, Progress in Optical Science and Photonics 23, https://doi.org/10.1007/978-981-99-0228-6_9

1.1 Bipolar OCDMA Code

In 2005 the base-band bipolar OCDMA system was introduced [1], where the transformation of unipolar–bipolar and bipolar-unipolar as well as the signal transmission process utilizing bipolar codes through unipolar medium are demonstrated. Gold sequence and m sequence as bipolar OCDMA code, as well as the effect of transmission medium on the performance of OCDMA system are elaborated here. The simulation results provided in the paper [1] specify that both the Gold and m sequence are redesigned to bipolar OCDMA systems.

Accordingly, the auto and cross-correlation performance shows that the Gold sequence is more desirable than the m sequence in the bipolar OCDMA system. One important scheme to transmit and detect the bipolar code sequences using a unipolar system is presented by Nguyen in 1995 [2]. The paper illustrates an all-optical implementation in non-coherent optical CDMA systems utilizing bipolar codes designed for the radio domain. The design described by Nguyen [2] shows the procedure to encode the broadband optical source spectrum to allow large number of subscribers.

A bipolar sequence $\mathrm{x_n}$ of period N is represented as the difference of two unipolar sequences having the same period. If one of the unipolar sequence is represent by u_n then the other is represented by $\overline{u_n}$ where u_n is derived from $\mathrm{x_n}$ by replacing each -1 with 0 and $\overline{u_n}$ is simply its complement [2]. The data symbols are encoded by u_n and $\overline{u_n}$ which modulate the optical signal source as on or off in two channels such that if symbol '1' is transmitted then u_n is on channel 1 and $\overline{u_n}$ on channel 2 and vice versa for symbol '0'. The received signal is decoded as Eqs. 1 and 2 as given below with the unipolar matched filters [2].

The two complementary correlator pairs produce output which are optically added followed by photo-detection. The balanced configuration of two photodiodes produces an outputs can be subtracted. The result is then passed through a low pass filter and compared to some threshold value to detect the output symbol as either '0' or '1'. Let '$\mathrm{v_n}$' is another unipolar sequence derived from bipolar code sequence $\mathrm{y_n}$ of period N. Therefore the periodic and aperiodic cross-correlation functions of $\mathrm{x_n}$ and $\mathrm{y_n}$ are expressed as follows.

$$\theta_{x,y}(l) = [\theta_{u,v}(l) + \theta_{\overline{u},\overline{v}}(l)] - [\theta_{u,\overline{v}}(l) + \theta_{\overline{u},v}(l)] \quad (1)$$

$$C_{x,y}(l) = [C_{u,v}(l) + C_{\overline{u},\overline{v}}(l)] - [C_{u,\overline{v}}(l) + C_{\overline{u},v}(l)] \quad (2)$$

Here the two correlator pairs are identical and the received signals are differentially decoded with one unipolar correlator pair and one photodetector then followed by a delay and a threshold comparator finally. The paper by Yen in 2013 reveals the study to analyze and calculate the dual unipolar and bipolar coded configurations of spectral amplitude coding (SAC) -OCDMA systems by simulation methods [3]. The significant attribute of the SAC-OCDMA systems is that the effect of MAI can be easily eliminated applying code sequences with a fixed in-phase cross-correlation (IPCC) value. The balance detection schemes are used to effectively cancel multiple access interference. In this paper [3] for unipolar and bipolar schemes Walsh-Hadamard signature codes are utilized. Optical Fiber Bragg gratings (FBGs) base encoder and decoder structures are used.

The simulated result shows that for a large received power the BER applying bipolar coding method is superior to the unipolar scheme. To transmit multimedia data a system with good performance can utilize the bipolar scheme in optical network. To transmit voice data the unipolar method is used. The eye diagram for bipolar encoding structure provides wider opening compared to unipolar encoding structure. A flexible code and simple hardware design with FBGs used to realize dual coding OCDMA system are suggested [3].

1.1.1 Walshcode

Walsh codes are mutually orthogonal error correcting codes and these codes posses interesting mathematical properties with a wide ranges of applications in communication systems. To enhance the bandwidth efficiency of the wireless CDMA Walsh code is used as it gives zero cross-correlation factor when properly synchronized in time [4]. Walsh code is described by N sequences set.

The length N where $N = 2^{n}$', and Walsh matrix is represented as $W_N^{(j)}$; when n are positive integers and $j \in [0, N-1]$ denotes the jth row of bipolar sequence derived from the W_N code set.

The Walsh code is a linear and mutually orthogonal code, binary strings of length n to create a binary code-word of length 2^n.

For the nth ordered $n \times n$ Hadamard matrix H contains 1 s and −1 s;

$HH^T = nI_n$ where $I_n = n \times n$ identity matrix [4].

To generate the Walsh codes, Hadamard matrix of the order 2^n is used. Hadamard matrices are possible for all orders for multiples of 4. Therefore, the WALSH codes [4] are generated by recursive procedure as follows:

$$W_1 = [+1]$$

$$W_2 = \begin{bmatrix} +1 & +1 \\ +1 & -1 \end{bmatrix}$$

$$W_4 = \begin{bmatrix} +1 & +1 & +1 & +1 \\ +1 & -1 & +1 & -1 \\ +1 & +1 & -1 & -1 \\ +1 & -1 & -1 & +1 \end{bmatrix} \text{And } W_{2^{n+1}} = \begin{bmatrix} W_{2^n} & W_{2^n} \\ W_{2^n} & \overline{W}_{2^n} \end{bmatrix}$$

where $\overline{W}_{2^n}$ represents the complement of W_{2^n}.

Hamming Distance of WALSH code: It is defined as the difference of bits for two different code words having length 'n' and for Walsh code the hamming distance is 2^{n-1} [2, 4].

1.1.2 Gold Sequence Code

Gold sequence is constructed using two Maximal Length sequences which can be simply designed by modulo 2 addition or XOR operation. WALSH codes are mainly developed to provide synchronous operations whereas maximal length sequences are used for code synchronization purpose. In case of asynchronous OCDMA system Gold sequences exist which is a family of bipolar code. It has very large cardinality along with good periodic cross-correlation property used to support synchronous operations [1].

A set of N + 2 Gold sequence having length $N = 2^{m}-1$ is designed from a pair of maximal-length sequences.

If x, y are two maximal-length sequences then the gold sequences set is provided as,

$$S_{gold} = \left[xyx \oplus yxy^{-1}x \oplus y^{-2} \ldots\ldots x \oplus y^{-i} \ldots\ldots x \oplus y^{(-(N-1))}\right]^T$$
$$Where-> y^{(-i)} = [y_i y_{i+1} \ldots\ldots y_{N-1} y_{N-1} y_0 y_1 \ldots\ldots y_{i-1}] \tag{3}$$

where $y^{(-i)}$ is the ith cyclic left-shift of y. These gold sequence possess a three valued cross-correlation function given as $-t_m$, -1 and t_m-2, where $t_m = 1 + 2^{(m+1)/2}$ when m is the odd number and for even m $t_m = 1 + 2^{(m+2)/2}$.

1.2 *Unipolar OCDMA Codes*

Unipolar codes are used for the traditional optical encryption and decryption process in case of incoherent OCDMA whereas the bipolar codes which are used in RF CDMA give poor performance for the OCDMA system. Hence the construction of the unipolar code with a good correlation property is needed for better system performance. Hence, unipolar codes need to be developed to improve the system performance. Evidently in the earlier years, research on incoherent OCDMA is aimed to obtain unipolar codes with good auto-and cross-correlation property. Many research works are motivated to develop an optical code having proper correlation values to provide better impact on interference reduction schemes. It can also be observed that a variety of unipolar one-dimensional codes are proposed and designed to employ in the OCDMA system.

The following section elaborates few important unipolar OCDMA codes

1.2.1 Prime Codes (PC)

In 1981, Titlebaum constructed the time–frequency hop code i.e., Prime code. These codes are mostly applicable to coherent multiuser Radar and asynchronous frequency hopping/spread spectrum communications. The construction of the code depends on linear congruence. The generation of the prime code structure is comparatively simple. There are two steps involved in designing the prime codes [5].

1st step: For a prime number *p*, a prime sequence $S_i = (s_{i,0}, s_{i,1}, \ldots\ldots s_{i,j}, \ldots\ldots\ldots, s_{i,p-1})$ is generated using the elements $S_{i,j} = i.j(\text{mod p})$, where, i and j all are in Galois field,GF(p) [5]. For example the prime sequence over GF(5) are $S_0 = (0,0,0,0,0)$, $S_1 = (0,1,2,3,4)$, $S_2 = (0,2,4,1,3)$, $S_3 = (0,3,1,4,2)$ and $S_4 = (0,4,3,2,1)$. 2nd step: Each of the prime sequence is portrayed into a binary code sequence as follows $CX = (CX_0, CX_1 \ldots\ldots CX_j \ldots\ldots CX_{(P^2-1)})$ according to the rule [5]

$$C_x = \begin{Bmatrix} 1, fori = S_x^p(j) + jp \bmod (p); j = 0, 1, \ldots\ldots p-1 \\ 0, otherwise \end{Bmatrix}$$

$$where, i = 0, 1, \ldots\ldots p^2 - 1 \quad (4)$$

As p number of ones exist within individual code, the cardinality of the prime code is p for GF(p).

The binary code sequences for GF(5) are:

$C_0{}^P = \{10{,}000\ 10{,}000\ 10{,}000\ 10{,}000\ 10{,}000\} = (0,0,0,0,0)$
$C_1{}^P = \{10{,}000\ 01{,}000\ 00{,}100\ 00{,}010\ 00{,}001\} = (0,1,2,3,4)$
$C_2{}^P = \{10{,}000\ 00{,}100\ 00{,}001\ 01{,}000\ 00{,}010\} = (0,2,4,1,3)$
$C_3{}^P = \{10{,}000\ 00{,}010\ 01{,}000\ 00{,}001\ 00{,}100\} = (0,3,1,4,2)$
$C_4{}^P = \{10{,}000\ 00{,}001\ 00{,}010\ 00{,}100\ 01{,}000\} = (0,4,3,2,1)$

From the above example it is found that a prime sequence set for code length and weight p^2 and p respectively supports only p distinct sequences. The Prime codes are derived from a GF(p), p = prime number. The Prime codes autocorrelation peak is at p, and the cross correlation value is always less than two. The correlation properties determine the performance of the code and in order to support a moderate number of subscribers it srequire much long code sequences.

The main limitation associated with the prime code is the limited availability of distinct sequences which results in limited number of possible subscribers. Also, the code length of the prime code is only P^2, and this will deteriorate the system performance in terms of BER hence MAI. Therefore, longer codes that maintain the desired properties are advantageous.

1.2.2 Modified Prime Codes (MPC)

Modified prime codes were mainly proposed to overcome the drawbacks of the prime codes [6]. The modified prime code could be generated by P−1 times shifting of the previous PC sequences. The code is capable of accommodating P^2 number of subscribers with P groups containing P sequences with code length that is equal to P^2, where P is a prime number [7]. The MPCs are generated by shifting left or right the original PC sequence. Then new time shifted sequences thus obtained are represented as $S_{X,t} = (S_{Xt0}, S_{Xt1}, \ldots\ldots S_{Xt(P-1)})$, where t represents the number of times S_X has been left (or right) rotated. Then the binary code sequence is generated according to the rule [7] stated below,

$$C_x = \begin{cases} 1, for, i = S_{stj} + jP, j = 0, 1, \ldots\ldots P - 1 \\ 0, otherwise \end{cases}$$

MPC with P = 5 is shown below in Table 2 [7]. Here MPC sequence length is P^2 and code weight P generates different code sequences P^2. Thus for an OCDMA system with MPC, P^2 number of possible subscribers can be allocated, which is much better with respect to PC, but the code length of MPC [7] is the same as PC [5] (Table 1).

To enhance system performance in terms of BER the user information is modified utilizing these designed codes for an incoherent OCDMA system. Again in the year 1993, Maric suggested a new class of optical code sequences which are utilized in spread-spectrum fiber-optic local area networks [9].

1.2.3 Quasi Prime Optical Orthogonal Code (OOC)

This code is the extension of the Optical Orthogonal Code (OOC) set and constructed from prime sequences as explained and elaborately by Holmes and Syms [6]. The quasi prime code $C_{xk}{}^{qp}$ is the time shifted and expanded or contracted version of the prime sequence code $C_x{}^p$.

For q number of ones the sequence is expressed as $c_{xk}{}^{qp}(i) = c_x{}^p([i + kp]_n)$;

Where i = 0,1,.....qp−1, n = qp, $(r-1)p < q < rp$;

p = prime number, q, r and k = positive integers;

The code sequence weight w = q. The auto correlation parameter $\lambda_a = (p-1)r$ and cross correlation parameter $\lambda_c = 2$, code-words number N is equal top[6].

For example for p = 5, q = 7, k = 3, the extended version of the prime codes are as follows:

$C_0{}^{qp}{}_k$ = {10,000 10,000 10,000 10,000 10,000 10,000 10,000}
$C_1{}^{qp}{}_k$ = {00,010 00,001 10,000 01,000 00,100 00,010 00,001}
$C_2{}^{qp}{}_k$ = {01,000 00,010 10,000 00,100 00,001 01,000 00,010}
$C_3{}^{qp}{}_k$ = {00,001 00,100 10,000 00,010 01,000 00,001 00,100}
$C_4{}^{qp}{}_k$ = {00,100 01,000 10,000 00,001 00,010 00,100 01,000}

Table 1 Modified prime sequence codes (P = 5) [7]

Code	Sequence patterns
MPSC {1, 1}	10,000 10,000 10,000 10,000 10,000
MPSC {1, 2}	01,000 01,000 01,000 01,000 01,000
MPSC {1, 3}	00,100 00,100 00,100 00,100 00,100
MPSC {1, 4}	00,010 00,010 00,010 00,010 00,010
MPSC {1, 5}	00,001 00,001 00,001 00,001 00,001
MPSC {2, 1}	10,000 01,000 00,100 00,010 00,001
MPSC {2, 2}	01,000 00,100 00,010 00,001 10,000
MPSC {2, 3}	00,100 00,010 00,001 10,000 01,000
MPSC {2, 4}	00,010 00,001 10,000 01,000 00,100
MPSC {2, 5}	00,001 10,000 01,000 00,100 00,010
MPSC {3, 1}	10,000 00,100 00,001 01,000 00,010
MPSC {3, 2}	01,000 00,010 10,000 00,100 00,001
MPSC {3, 3}	00,100 00,001 01,000 00,010 10,000
MPSC {3, 4}	00,010 10,000 00,100 00,001 01,000
MPSC {3, 5}	00,001 01,000 10,000 00,010 00,100
MPSC {4, 1}	10,000 00,010 01,000 00,001 00,100
MPSC {4, 2}	01,000 00,001 00,100 10,000 00,010
MPSC {4, 3}	00,100 10,000 00,010 01,000 00,001
MPSC {4, 4}	00,010 01,000 00,001 00,100 10,000
MPSC {4, 5}	00,001 00,100 10,000 00,010 01,000
MPSC {5, 1}	10,000 00,001 00,010 00,100 01,000
MPSC {5, 2}	01,000 10,000 00,001 00,010 00,100
MPSC {5, 3}	00,100 01,000 10,000 00,001 00,010
MPSC {5, 4}	00,010 00,100 01,000 10,000 00,001
MPSC {5, 5}	00,001 00,010 00,100 01,000 10,000

Table 2 Example of MFH codes [8]

α	β	Sequences	Sequence of binary numbers (MFH codes)
0	0	12,300	0100 0010 0001 1000 1000
1	0	23,101	0010 0001 0100 1000 0100
2	1	20,312	0010 1000 0001 0100 0010
0	3	21,030	0010 0100 1000 0001 1000
2	2	13,022	0100 0001 1000 0010 0010

1.2.4 Quadratic Congruence Codes (QCC)

Each Quadratic Congruence Code sequence of weight p and length $N = p^2$ gives a cardinality p−1. For the code construction a quadratic placement operator y(m) is defined by Marie [9, 10] as follows:

$$S_{i,j} = \frac{i.j.(j+1)}{2}(\mathrm{mod}\, p),\ \text{for } 0 \le j \le p-1 \text{ and } 1 \le i \le p-1$$

For any user k the code sequence, $c_k = \{c_k(i)\}$ for i = 0 to N− is constructed. The sequences are of length $N = p^2$ and weight p. There can be at most (p-1) different code sequences that can be constructed for each p. the maximum auto-correlation and cross-correlation value is 2 and 4, respectively. Hence QCC are characterized by the quadruple (p^2, p, 2, 4).

1.2.5 Modified Quadratic Congruence (MQC) Codes

In this paper, Z. Wei suggested the generation process for a series of new code families with a fixed in-phase cross correlation value of 1. To neglect PIIN's effect, based on block designs theorem a new code established by $\frac{q^{m+1}-1}{q-1}, \frac{q^m-1}{q-1}, \frac{q^{m-1}-1}{q-1}$ has been introduced [11]. It is observed that higher SNR can be achieved by effectively eliminating the intensity noise. The higher SNR is the result of higher ratio of the autocorrelation peak to the fixed in-phase cross correlation. With MQC, for each prime number p, one gets a series of new codes $[p^2 + p, p + 1, 1]$.

It is observed that these new codes have similar properties with some other few existing important codes. The design of both the MQC based transmitter and receiver configuration for the SAC-OCDMA system is provided using FBG groups. The comparison of BER performances with the Hadamard code is also presented. This paper work proved that when both the PIIN and shot noise are taken into account, the count number of the arrived incoming photons obeys a negative binomial distribution [11]. To reduce the PIIN without utilizing too many FBGs, MQC code can be used [11, 12].

For one active user, the negative- binomial-distributed photoelectron counting produces a small bit error rate (BER). This result is also experimentally verified by Dennis and Young [13].

1.2.6 Modified Frequency Hopping (MFH) Codes

Modified frequency hopping (MFH) is a family of frequency hopping (FH) codes with ideal cross-correlation based on the main power. It was originally introduced in (2002) by Wei and Ghafouri [8]. Therefore, MFH is characterized by small code length N = q2 + q, code weight W = q + 1 with ideal cross-correlation $\lambda c = 1$ but code design is complex. Listed below are the code words for various values for the parameters α and β when $q = p^n = 2^2$, where p is the prime number with n being a positive integer, $\alpha \in \{0, 1, 2,..., q-2\}$ and $\beta \in \{0, 1, 2,...,q-1\}$.

The MFH code can exist for a much wider number of integers than the MQC code, because it depends on the power of the prime number q. Again like the MQC code, the MFH code provides q^2 sequences. Each of the MFH code pattern has $q^2 + q$ elements separated into (q + 1) groups and each group holds one '1' and (q−1) '0's.

1.2.7 Random Diagonal (RD) Codes

In 2010, Fadhil and his co-authors proposed the Random Diagonal (RD) code for the SAC-OCDMA system [13, 14]. Code level and data level are used to design the RD code. One of the important features of this code is that the data-level cross-correlation is zero. The code length necessitates to be increased to support a greater number of users. Here it is essential to repeat each row on both matrices [M] and [B]. For the Kth user matrix, [M] and [B] are expressed as:

$$[M] = \begin{bmatrix} 0 & 1 \\ 1 & 0 \\ 0 & 1 \\ 1 & 0 \\ 0 & 1 \\ & \\ & \\ a_{j1} & a_{j2} \end{bmatrix}$$

where j is the value for Kth user (j = 1,2…K), and a_j value is either zero or one.

For a basic code matrix Z the user number is K with code length N. Z is represented by (K × N) matrix [13, 15]. Therefore it is observed that the code weight, i.e., the number of ones in each row is equal to 4. The relation between N and K for W = 4 is N = K + 5.

$$[Z] = \begin{bmatrix} 0\,0\,1\,1\,1\,0\,1\,0 \\ 0\,1\,0\,0\,1\,1\,0\,1 \\ 1\,0\,0\,1\,0\,1\,1\,0 \end{bmatrix}$$

To conclude, the advantages of RD codes are shorter code length, cross-correlation value is zero for the data level which minimized λ and reduced PIIN. It also provides flexible selection of N, K parameters compared to MFH and MDW codes.

System performance with RD code is analyzed considering the impact of intensity noise, shot and thermal noise sources. It is observed that the RD code performs best in terms of BER compared to the MFH code.

1.2.8 Enhanced Double Weight (EDW) Codes

A new code structure for the SAC-OCDMA system utilizing double weight (DW) code families is suggested by Hasoon [8, 15]. Enhanced double-weight (EDW) code with variable weight (>1) is another variation of a DW code (weight is fixed at 2) and it also possesses ideal cross-correlation properties. Enhancement of system performance is achieved utilizing the EDW code compared that Hadamard [12] and Modified Frequency-Hopping (MFH) [10] codes.

The DW code construction uses the following steps [8, 16]:

Step 1:The DW codes are represented with K X N matrix.

Where K = No of rows = user number, and N = No. of columns = minimum code length

The basic DW code with K = 2 and N = 3 is shown below:

$H_1 = \begin{bmatrix} 0\,1\,1 \\ 1\,1\,0 \end{bmatrix}$, We can see that the H_1 has a chips combination sequence of 1, 2, 1 for the three columns (i.e., 0 + 1, 1 + 1, 1 + 0).

Step 2:A simple mapping technique is used to increase the number of codes as shown below:

$$H_2 = \begin{vmatrix} 0 & H_1 \\ H_1 & 0 \end{vmatrix}$$

EDW code construction: The EDW code is the enhanced version of the DW code. The code weight is any odd number greater than one. In reference [8, 15] the EDW code with a weight equal to three is elaborated. The code can be created using the steps below.

Step 1: like DW, the EDW code is also represented by a simple K x N matrix. A basic EDW code for K = 3 and N = 6 is shown below [8, 15]:

$$H_0 = \left|\begin{array}{c|c|c} 0\,0 & 1\,1 & 0\,1 \\ 0\,1 & 0\,0 & 1\,1 \\ 1\,1 & 0\,1 & 0\,0 \end{array}\right|$$

It can be observed that the structure of the EDW code maintained quite similarities with the DW code matrix with slight variations such as double weight pairs held in such a way that only two overlapping chips are allowed in each column.

Step 2: For a larger number of K a mapping process is used to generate H_1 as follows.

$$H_1 = \begin{matrix} 0 & H_0 \\ H_0 & 0 \end{matrix}$$

An EDW code with weight 3 is presented by (N, 3, 1) for code length N. The relation among code-length weight and user number is presented as [8, 15]

$$N = 2K + \frac{4}{3}\left[\sin\left(\frac{K\pi}{3}\right)\right]^2 \frac{8}{3}\left[\sin\frac{(K+1)\pi}{3}\right]^3 + \frac{4}{3}\left[\sin\frac{(K+2)\pi}{3}\right]^2 \quad (5)$$

For EDW codes [8, 15], the weight can be set to any odd number regardless of the number of users, resulting in a simplified encoder/decoder design. The SNR is not affected by the number of users, therefore the same quality of service can be provided to all users. One of the main advantages of the EDW code is that it offers improved SNR than the MFH and Hadamard codes.

This can be seen from the fact that the EDW code provides ideal cross-correlation, while the Hadamard code has an increasing value of cross-correlation as the number of user increases. The cross-correlation value is also fixed to one in the case of MFH codes since the SNR value is smaller than that of EDW.

1.2.9 Modified Double-Weight (MDW) Codes

The MDW code is in turn another revised version of the DW code and was proposed by Aljunid and his co-authors in 2004. Unlike the DW code, the redesigned code has variable weights. Since the basic DW code contains K rows for user number and N columns for minimum code length represented as [15, 16]

$$H_1 = \begin{bmatrix} 0\,1\,1 \\ 1\,1\,0 \end{bmatrix}$$

Now to construct the MDW code of basic matrix (9, 4, 1) a simple mapping technique has been employed by the author as below

$$H_{m1} = \left(\begin{array}{c|c|c} 0\,0\,0 & 0\,1\,1 & 0\,1\,1 \\ 0\,1\,1 & 0\,0\,0 & 1\,1\,0 \\ 1\,1\,0 & 1\,1\,0 & 0\,0\,0 \end{array}\right)$$

The supported number of users K for the MDW code related to N is given by Equation.

$$N = 3K + \frac{8}{3}\left[\sin\left(\frac{K\pi}{3}\right)\right]^2 \quad (6)$$

The table below points out some basic properties of the afore-mentioned codes (Table 3).

Table 3 Properties of various OCDMA codes

Code type	Code-length N	Code weight W	Cross-correlation value, λc	Observation(s)
MQC [11]	$p^2 + p$, p = prime no	$p + 1$	1	The requirement of prime number limits its freedom of code selection
MFH [8]	$q^2 + q$, q = prime no	$q + 1$	1	Code construction and the design of encoder and decoder are complicated
HD [17]	N	2^{M-1}	2^{M-2}	Correlation factor is not ideal which leads to less effective to MAI
EDW [8]	$N = 2K + \frac{4}{3}\left[\sin\left(\frac{K\pi}{3}\right)\right]^2 \frac{8}{3}\left[\sin\frac{(K+1)\pi}{3}\right]^3 + \frac{4}{3}\left[\sin\frac{(K+2)\pi}{3}\right]^2$	3	1	Easy code construction as well as simple encoder decoder architecture
RD [13]	K + 2w-3	≥ 4	Variable	Simple code construction and more immune to MAI
MDW [15]	$N = 3K + \frac{8}{3}\left[\sin\left(\frac{K\pi}{3}\right)\right]^2$	Any even no greater than 2	1	More immune to MAI

1.3 *Weighted Position Representation (WPR) of Unipolar OCDMA Code (1-D)*

The 1-D unipolar OOC word X with code length n and code weight w contains w number of ones and $(n\text{-}w)$ number of zeroes. For code length n there are n possible positions ranges from 0th to $(n\text{-}1)^{th}$ places where exists exactly w weighted places and $n\text{-}w$ non-weighted places. The code X is denoted simply with its weighted positions. It possible to derive n presentations for each of n circular shifted versions of the code X.

This kind of depiction of the unipolar codeword is called weighted positions representation (WPR) or bit 1's position representation [15, 16, 18]. As an example for the code X of length $n = 11$, code weight $w = 5$so as X = 10,100,101,001 the WPR is (0,2,5,7,10). Any n circular shift of code X expresses same unipolar orthogonal code X.

There are also several OOCs which are based on the difference set and projective geometry too which we have not considered in this chapter.

1.4 Correlation Constraints of Unipolar Optical Codes

For OCDMA system, one bit is fragmented into '*n*' time periods, defined as "*chips*". A short optical pulse is transmitted in the chip intervals to generate any signature sequence. The entire number of illuminated chips within the code sequence is defined as Hamming weight *w*. In case of the optical CDMA system each user can be represented by its unique signature sequence called the *codeword.*

The encoder at the transmitter end encodes each '1' utilizing the signature sequence and incase of '0' data bit encoding is not required and presented using an all-zero sequence. The encoded signal is than transmitted to the *N* x *N* star coupler which is next broadcasted to every node. The dominant noise source in the OCDMA system is multiple-access interference (MAI) caused by the crosstalk among different users sharing the same fiber optic channel. Therefore, smart code sequences design is important to reduce or nullify MAI effect from the received signal. Let the two unipolar code words *X* and *Y be* within the code set with the code parameters (n, w, λ_a, λ_c). The autocorrelation Property is given by [19–21]:

$$\sum_{t=0}^{n-1} x_t x_{t+\tau} \leq \lambda_a; \text{For } x \in C \text{ and for any integer } \tau \text{ such that } 0 < \tau < \text{n} \quad (6)$$

The cross-correlation property,

$$\sum_{t=0}^{n-1} x_t x_{t+\tau} \leq \lambda_c; \text{ for any integer } x \neq y \in C \text{ and any integer} \tau \quad (7)$$

Here C is the code. The (0, 1) sequences of an OOC are the codeword. The correlation properties are not affected by cyclic shifts of the code-words. The code C is considered as a family of w-sets of integers $|n|$. Then the correlation property is re-formulated as below [19].

$$|(a + X) \cap (b + X)| \leq \lambda_a$$

For any $X \in C$ and any $a \neq b(modn)$.

The cross-correlation property: $|(a + X) \cap (b + Y)| \leq \lambda_c$

For $X \neq Y \in C$ and for any a, b where a + X = {a + x:$x \in X$}

An upper bound to the maximum size of code is denoted as the cardinality factor ϕ(n,w,λ) [19–21]. For an C(n,w, λa, λc) is deifned as a collection of binary n-tuples, each having weight w.

Techhniques of error control codes are used to construct OOCs with λa, λc as small as possible. For optimum OOCs the modified upper bound derived from Johnson bound to get error free codes is obtained for λa = λc = λ as cited in [19].

$$C_{1D} = \varphi(N, w, \lambda) = \frac{(N-1)(N-2)....(N-\lambda)}{w(w-1)(w-2).....(w-\lambda)} \quad (8)$$

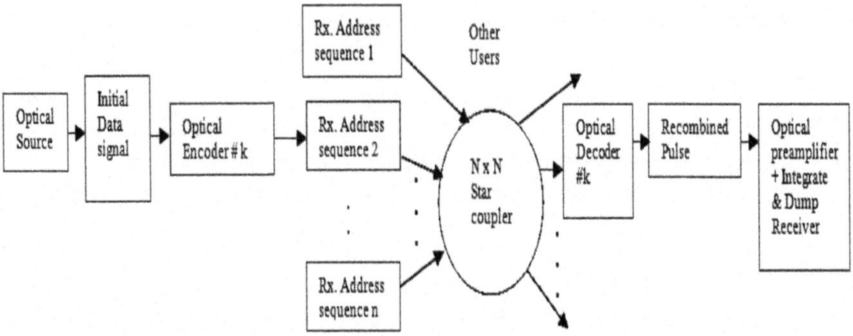

Fig. 1 Transmitter–receiver architecture for fiber-optic-CDMA systems [20, 22, 23]

1.5 Review of Different Transmitter Receiver Architectures of the OCDMA System

This section gives an overview of a few well known transmitter receiver structures designed for user information detection schemes and to improve the system performance and increase the capacity by reducing the multi access interference (MAI) effect.

1.5.1 Basic OCDMA System Structures for Fiber-Optic-CDMA System

Basic OCDMA system Structures for Fiber-Optic-CDMA Systems: In 2007 Salehi [20] suggested the working principle and detailed description of the receiver structure based on star configuration. The proposed architecture contains N transmitter and receiver pairs. In this case every user's information bit is mapped into a high rated optical sequence by an optical encoder. At the receiver section the optical pulse is then compared to already stored data same as the data used in the transmitter followed by threshold comparator to detect the original data. The optical match-filter used contains w fiber delay lines to produce a delay equal to (L − c1)Tc, (L − c2)Tc,..., and (L − cw)Tc respectively (Fig. 1).

1.5.2 OCDMA System Architecture Using Correlator Receiver Structure

For a correlator receiver architecture (Fig. 2) the initial modification suggested by Salehi and Brackett [21]. The authors observed that at a particular chip time the amount of intensity transmitted for a '1' bit may carry information and any excess intensity owing to interference. Such interference may cause a '0' data bit to be wrongly received as '1' or vice-a-versa.

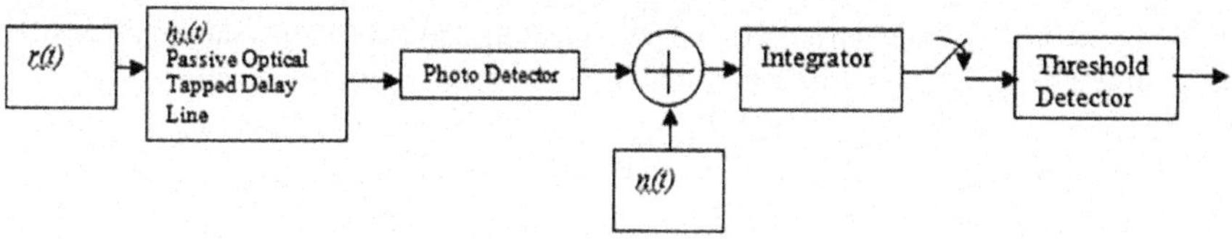

Fig. 2 Architecture of correlator receiver [21]

Thus to limit the unwanted amount of intensity an optical hard-limiter is placed followed by the traditional correlator receiver which in-turn was able to block some interfering patterns for enhancement of system performance. To suppress interference causing some error patterns the authors suggested employment of an optical hard-limiter for improving system performance against multi user interference (MUI) or MAI.

1.5.3 Transmitter–Receiver Architecture Using Modified-AND Subtraction Detection

The optical signal detection technique using Modified-AND subtraction was suggested by Hamza and his co-authors [22] in the year 2012. From their work it is perceived that due to incoherent broadband light sources SAC-OCDMA systems are more prone to intensity noise. The detection technique employing the modified-AND subtraction provides a scope to suppress this noise in incoherent SAC-OCDMA systems.

The detection mechanism relies on lowering the received signal strength during the decoding process by the spectrum division of the employed code sequence. The SAC-OCDMA receiver structure using the proposed method is shown (Fig. 3).

In this configuration the received optical power is divided into two parts; one provided to the upper part of the decoder and the other to the complementary decoder

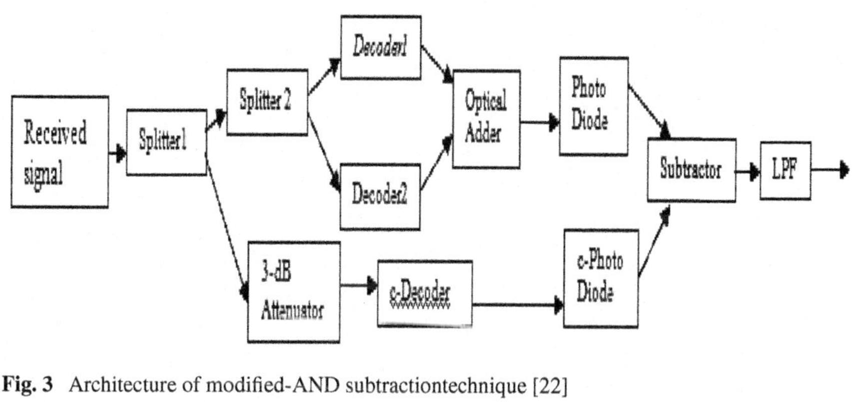

Fig. 3 Architecture of modified-AND subtractiontechnique [22]

utilizing attenuator to ensure that interference signal produces the same power incident on each of the photo-detector when the desired user is transmitting '0'. The signal accepted at the decoder is half of the incident signal processed by the conventional AND subtraction detection process. This decoding technique uses fiber Bragg-gratings (FBGs) due to low insertion losses, better spectral resolution, compressed size and uses of economical material.

Various filtering techniques are there for decoding signal like FBGs, arrayed waveguide-gratings (AWGs) etc. In this case the cancellation of significant interference can be achieved by the difference of the two signals. The original data are restored after passing via a decision circuit.

1.5.4 Transmitter–Receiver Architecture Using NAND Subtraction Detection Technique

Ahmed and his co-authors in the year 2012 conceptualized the detection technique using the NAND subtraction process based on the Modified Double Weight (MDW) code [15] for SAC-OCDMA systems.

In this method, the concept of NAND is utilized as an operation instead of the digital gate to recover data. Here for the detection of the output they replaced the cross-correlation by the NAND operation between two sequences X and Y. Figure 4 above represents the detection mechanism. But, a drawback of this technique is that it may generate extra weight due to which the signal power increases as well as the code weight.

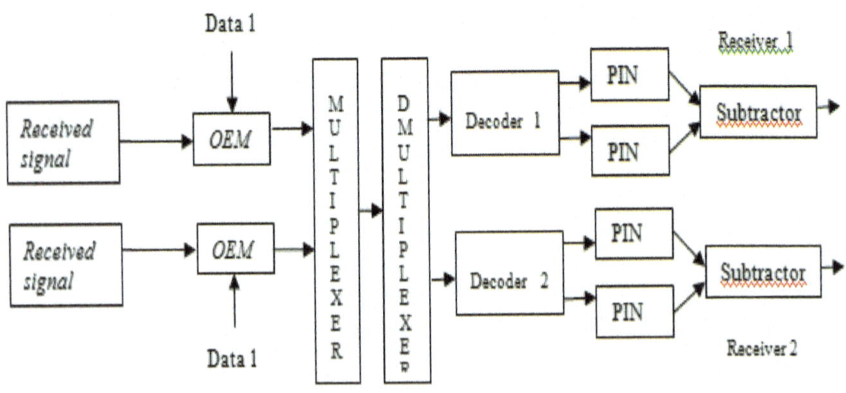

Fig. 4 Block diagram of implementation of NAND subtraction detection technique [22]

1.5.5 Structure of Transmitter and Receiver Based on FBG

In 1997, Kenneth O. Hill and Gerald Meltz described properties and the fundamental properties of fiber gratings (FBG) [24]. Here they showed that the index perturbation in the core is a periodic structure similar to a crystal lattice that acts as a stop-band filter. Applications of FBGs are extensive in optical networks such as add-drop multiplexing and wavelength routing in WDM networks [24–26]. FBGs are also used to design various OCDMA coding schemes such as B. spectral amplitude coding, time phase coding and spectral phase coding.

FBGs offer a simple and elegant all-fiber approach to provide tunability using strain and temperature effects. The implementation of WHTS using FBGs is done using two structures: a linear array of FBGs and Chirped Moiré Gratings (CMG). Figure 5 below shows the implementation of a WHTS code using a linear array of FBGs [21].

From Fig. 5b it can be observed that for a specific column of the code matrix the maximum number of wavelengths is 'k'. The chip size in FBG-based WHTS encoder-decoders is restricted by adjacent gratings minimum spacing [25]. The disadvantage of this kind of system is the requirement of further components like power splitters or circulators to separate the coded decoded signals which leads to increase in system complexity and cost (Fig. 6).

The spectral components proportional to the spectral distribution $A(v)$ are reflected back when a broadband pulse is injected into a group of FBGs, and the output at the other end of the grating contains all the complementary components proportional to $\overline{A(v)}$. At the receiver end the output from the top of the initial FBG group is directly used as to decode the output.

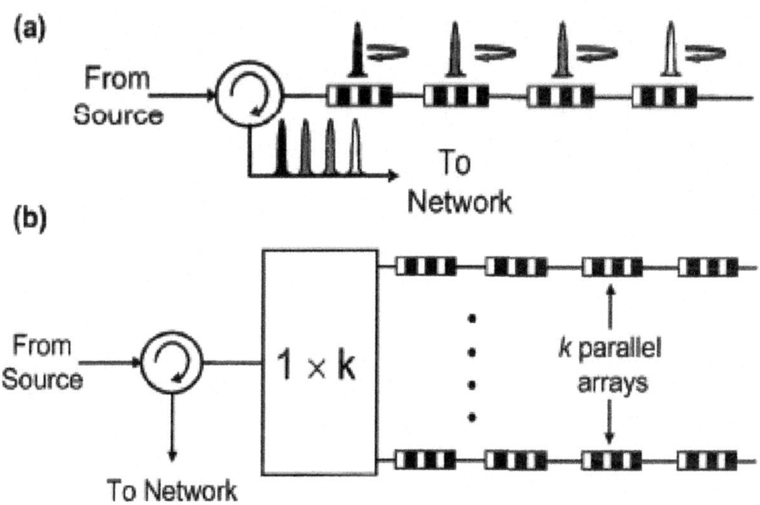

Fig. 5 **a** Schematic of a WHTS encoder with four wavelengths using a linear array of FBG. **b** Parallel linear array of FBGs allows implementation of OCDMA codes [21]

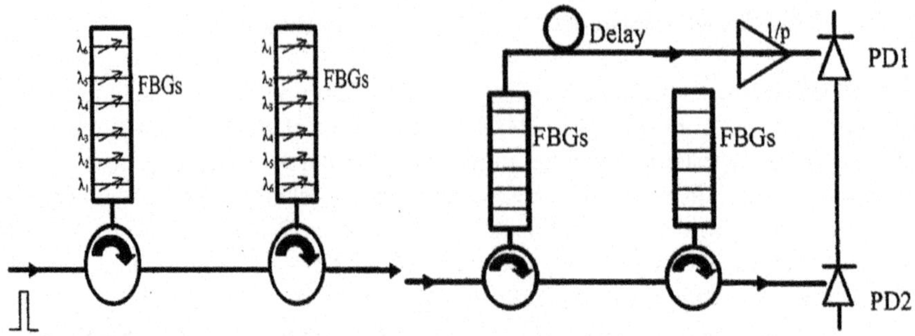

Fig. 6 Encoder decoder using FBG [25]

1.5.6 AWG Based OCDMA Encoding–Decoding Technique

Since the early 1990s, AWG has been of great research interest due to its low insertion loss [25]. As shown in the figure below, it contains a phased array of optical waveguides used as a grating. After passing through a region of free propagation, the input signal is received in an array of planar waveguides (Fig. 7).

The waveguides are designed in such a fashion that their lengths result in different phase shifts to the signal in each waveguide. Owing to the dependence on frequency of the mode propagation constant these phase shifts are wavelength dependent. AWGs has wide range of applications like wavelength multiplexers or demultiplexers, N × N wavelength routers [25], optical cross connects and so on. AWGs with more than 100 ports are presented in [21]. In general commercially available devices have more than or equal to 50 ports. For the WHTS optical encoder/decoder structure AWG can be combined with the waveguide-based delays.

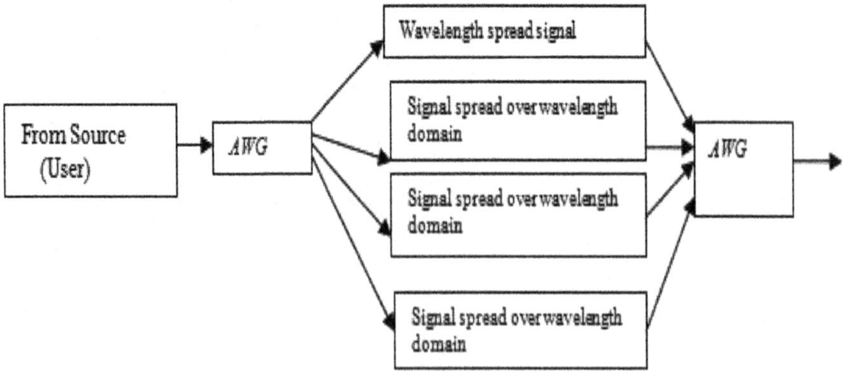

Fig. 7 Block diagram of for implementation of a WHTS encoder/decoder using AWG [25]

2 Conclusion

This chapter explains the knowledge of few important as well as well-known bipolar and unipolar optical orthogonal codes used in the OCDMA system. Properties of these bipoalr and uniploar codes have also been highlighted. The conventional representation and conventional methods for calculation of auto-correlation constraint and cross correlation constraints are discussed as well. The chapter also elaborates different transmission and detection schemes suitable for the non-coherent OCDMA system.

From the above discussion we can conclude that there are various encoding and decoding schemes which are proposed to minimize the MUI or MAI to achieve better system performance. For an OCDMA system the communication between each pair of transmitter and receiver is implemented by encoding and decoding data with a unique code sequence. To achieve this, perfect synchronization between the desired transmitter and receiver pair is essential. Hence the transmitter and receiver are all required to be well-tuned.

References

1. R. Adams, J. Faucher, L. Thomas, L.R. Chen, Demonstration of encoding and decoding 2-D wavelength-time bipolar codes for OCDMA systems with differential detection. IEEE Photonics Technol. Lett. **17**(11), 2490–2492 (2005)
2. L. Nguyen, B. Aazhang, J.F. Young, All-optical CDMA with bipolar codes. Electron. Lett. **31**(6), 469–470 (1995)
3. C.T. Yen, H.C. Cheng, Y.T. Chang, W.B. Chen, Performance analysis of dual unipolar/bipolar spectral code in optical CDMA systems. J. Appl. Res. Technol. 11(2) (2013)
4. K. Singhal Walsh codes, PN sequences and their role in CDMA technology, in *Term Paper–EEL* (2012)
5. G.C. Yang, W.C. Kwong, Prime codes with applications to CDMA optical and wireless networks. (Norwood, Artech House, 2002)
6. A.S. Holmes, R.R.A. Syms, All-optical CDMA using "quasi-prime codes. J. Lightwave Technol. **10**, 279–286 (1992)
7. J.G. Zhang, W.C. Kwong, Effective design of optical code-division multiple access networks by using the modified prime code. IEE Electronics Letters **33**(3), 229–230 (1997)
8. F.N. Hasoon, S.A. Aljunid, M.K. Abdullah, S. Shaari, Spectral amplitude coding OCDMA systems using enhanced double weight code. J Eng Sci Technol **1**(2), 192–202 (2006)
9. S.V. Maric, Z.I. Kostic, E.L. Titlebaum, A new family of optical code sequences for use in spread-spectrum fiber-optic local area networks. IEEE Trans. Commun. **41**, 1217–1221 (1993)
10. S.V. Maric, A new family of optical code sequences for use in spread-spectrum fiber-optic local area networks. IEEE Trans. Commun. **41**(8), 1217–1221 (1993)
11. Z. Wei, H.M. Shalaby, H. Ghafouri-Shiraz, Modified quadratic congruence codes for fiber bragg-grating-based spectral amplitude-coding optical CDMA systems. J. Lightwave Technol. **19** (2001)
12. Z. Kostic, The design and performance analysis for several new classes of codes foroptical synchronous CDMA and for arbitrary-medium time-hopping synchronous CDMA systems. IEEE Trans. Commun. **42**(8), 2608–2617 (1994)
13. H.A. Fadhil, S.A. Aljunid, B. Ahmed, Performance of OCDMA systems using random diagonal code for different decoders architecture schemes. Int. Arab J. Inform. Technol. **7**(1) (2010)

14. S.A. Aljunid, A.R. Ramli, M.A. Borhanuddin, K.A. Mohamad, A new family of optical code sequences for spectral-amplitude-coding optical CDMA systems. IEEE Photon. Technol. Lett. **16**(10), 2383–2385 (2004)
15. N. Ahmed, S.A. Aljunid, H.A. Fadil, R.B. Ahmad, M.A. Rashid, Performance improvement of SAC-OCDMA system using modified double weight (MDW) code for optical access network. Sci. Res. Essays **7**(7), 796–804 (2012)
16. R.C.S. Chauhan, R. Asthana, Y.N. Singh, A general algorithm to design sets of allpossible one dimensional unipolar orthogonal codes of same code length and weight, in *2010 IEEE International Conference on Computational Intelligence and Computing Research (ICCIC-2010)*. (Coimbatore, India, IEEE, 2010), pp. 7–13
17. Z. Wei, H.G. Shiraz, Unipolar codes with ideal in-phase cross-correlation for spectral-amplitude-coding optical CDMA systems. J. Lightwave Technol. **50**(8) (2002)
18. F. Chung, J.A. Salehi, V. Wei, Optical orthogonal codes: design, analysis and applications. IEEE Trans. Inf. Theory **35**(3), 595–604 (1989)
19. P.R. Prucnal, Optical code division multiple access: fundamentals andapplications. (CRC Press, 2006)
20. J.A. Salehi, Emerging Optical CDMA Techniques and Applications (Invited Paper). Int. J. Optics Photon. (IJOP) **1**(1) (2007)
21. J.A. Salehi, C.A. Brackett, Code division multiple access techniques in optical fiber networks-Part II: Systems performance analysis. IEEE Trans. Commun. **37**(8), 824–833 (1989)
22. H.M. Al-Khafaji, S.A. Aljunid, H.A. Fadhil, Improved BER based on intensity noise alleviation using developed detection technique for incoherent SAC-OCDMA systems. J. Modern Optics (2012)
23. S. Sikder, S. Ghosh, Study of bipolar Walsh coding in optical code division multiple access (OCDMA) system in Proc. Nat. Conf. Mater. Devices Circuits Commun. Technol. (IETE Burdwan Sub Centre, University of Burdwan, India 2012), pp. 200–203
24. K.O. Hill, G. Meltz, Fiber bragg grating technology fundamentals and overview. J. Lightwave Technol. **15**(8), 1263–1276 (1997)
25. L.R. Chen, Flexible fiber Bragg grating encoder/decoder for hybrid wavelength-time optical CDMA. IEEE Photon. Technol. Lett. **13**(11), 1233–1235 (2001)
26. S. Ghosh, B.N. Chatterji, Theoretical study of use of optical orthogonal codes for compresses video transmission in optical code division multiple access (OCDMA) systems, in *Proceedings of International Conference on Photonics & optical Engineering,* (California, 2007)
27. T. Dennis, J.F. Young, Measurements of BER performance for bipolar encoding of an SFS. J. Lightwave Technol. **17**, 1542–1546 (1999)
28. R.C.S. Chauhan, R. Asthana, Y.N. Singh, Unipolar orthogonal codes: design, analysis and applications. *International Conference on High Performance Computing (HiPC-2010), Student Research Symposium,* (Goa, India, 2010)
29. H. Yin, D.J. Richardson, Optical code division multiple access communication networks theory and applications. (Springer, Tsinghua University Press Publication)
30. N. Ahmed, S.A. Aljunid, R.B. Ahmad, M.A. Rashid, Novel OCDMA detection technique based on modified double weight code for optical access network. ElektronikaIrElektrotechnika **18**(8), 117–121. ISSN 1392–1215 (2012)
31. R. Kashyap, *Fiber Bragg gratings* (Academic Press, San Diego, 2009)

Error Detection and Correction of High-Resolution Remote Sensing Images Using Cyclic Code

Anirban Patra, Debasish Chakraborty, S. K. Sohel Ahamed, Subhadeep Ghosh, and Santanu Kamilya

1 Introduction

For effective transmission of high-resolution images, the requirement of a secured system is the primary criterion. This security can be implemented using different coding and encryption processes. Cyclic coding is implemented to code medical images in Covid–19 pandemic situations. In this method, the selected medical images are coded using a generator polynomial maintaining a cyclic code algorithm [1]. In an optical domain system, by using sinusoidal phase and amplitude grating with the help of a random phase matrix, encryption, as well as compression, are performed [2–5]. In these systems, the selected images are modulated with the help of high grating frequency and variable orientation angles. As a result of the modulation, multiple spectra are generated which are encrypted using a random phase matrix. This system provides excellent results in ciphertext attacks. DES, symmetric and asymmetric algorithms are the few oldest methods used for encryption. The major advantage of the DES system is that it takes minimum time to encrypt the messages [6, 7].

Use of Cipher Feedback Mode enhanced the security of the medical image encryption process [8]. The encryption system is based on the (k, n) method, where by selecting the proper values of k and n, the selected images can be retrieved using the proper retrieval system [9]. This system is proposed by Shamir. By shuffling the pixel position and applying XORing function the on image encryption method was proposed by Nag et al. [10]. Since then, researchers applied different techniques for the encryption of images like a one-time password system [11] which divides the image bits into blocks of rows and columns, for example, The Dael algorithm process

A. Patra (✉) · S. K. S. Ahamed · S. Ghosh · S. Kamilya
JIS College of Engineering, Kalyani, West Bengal, India
e-mail: anitublu@gmail.com

D. Chakraborty
Regional Remote Sensing Centre (East), ISRO, Kolkata, West Bengal, India

A. Saha et al. (eds.), *Optical to Terahertz Engineering*, Progress in Optical Science and Photonics 23, https://doi.org/10.1007/978-981-99-0228-6_10

[12], the AES system [13], etc. Cyclic coding is used in binary and low-resolution image processing for encoding [14–17]. Recently Hamming code is used for error detection as well as correction [18] and grating is used to store more images [19].

In this communication, the selected images are encoded using the systematic cyclic code. Here, a well-known generator polynomial (3°) is utilized to encode the image. Initially, all the pixels of the selected images are converted into bits for proper processing. In the next phase, the converted pixels are encoded with the help of a generator polynomial to generate a codeword. In this way, all pixels are converted into code words. This encoded image is to be transmitted over a long distance. On receiving, it has been checked whether the received images are free from error or not. If the error is present in the images, with the help of this algorithm, the position of the error has been detected from the syndrome table. To eliminate errors from the images, we have to swap the binary value of this position. To measure the quality of the output images, the conventional PSNR (Peak Signal to Noise Ratio) method has been used. With the help of simulation software, we have proved that our method provides a satisfactory result.

2 Methodology

The entire process is divided into categories: codeword generation and error detection. During the codeword generation part, we have followed these processes:

Selected generator polynomial g(x)
Pixel of selected image f(x, y) is represented by p(x)

If n = codeword length and k is the message length

$$m(x) = p(x) \cdot X^{(n-k)} \tag{1}$$

To achieve the remainder r(x), m(x) is divided by the generator polynomial g(x). Hence,

$$r(x) = m(x)/g(x) \tag{2}$$

m(x) and r(x) are added together to generate code word c(x).

Therefore,

$$c(x) = m(x) + r(x) \tag{3}$$

This codeword c(x) is transmitted over a long distance. Let us assume that the received codeword is k(x).

At the receiving end, we have to divide the k(x) by the same generator polynomial g(x) to get the remainder r1.

Therefore,

Table 1 Syndrome table

Syndrome	Error bit position
0	No error
1	1st bit
x	2nd bit
x2	3rd bit
x + 1	4th bit
$x^2 + x$	5th bit
$x^2 + x + 1$	6th bit
$x^2 + 1$	7th bit

$$r1 = k(x)/g(x) \tag{4}$$

There should be no difference between c(x) and k(x). If there is no error (the same codeword is received), the value of r1 is zero. Otherwise, we have to check the position of the error from the syndrome table.

Let us assume that p(x) is 0110. It is represented by $x^2 + x + 1$. Generator polynomial g(x) is $x^3 + x + 1$ (which is 1011)

$$[n = 7 \text{ and } k = 4]$$

So, $m(x) = p(x).\ x^3 = (x^2 + x + 1).x^3 = x^5 + x^4 + x^3$

Remainder $r(x) = (x^5 + x^4 + x^3)/(x^3 + x + 1) = x$

Hence, $c(x) = m(x) + r(x) = x^5 + x^4 + x^3 + x$ which is equivalent to 0111010.

At the receiving end, we have to check the remainder r1(x). Remainder r1(x) is also known as syndrome. The position of the remainder value r1(x) and the respective error bit is available from the syndrome table (Table 1).

Mean Square Error (MSE) can be represented as,

$$MSE = \frac{1}{mn}\sum_{y=1}^{m}\sum_{x=1}^{n}[f(x, y) - g(x, y)]^2 \tag{5}$$

f(x, y) and g(x, y) represent the original image and output image respectively;

$$PSNR = 20\log 10\frac{255}{\sqrt{MSE}}. \tag{6}$$

3 Flow Chart

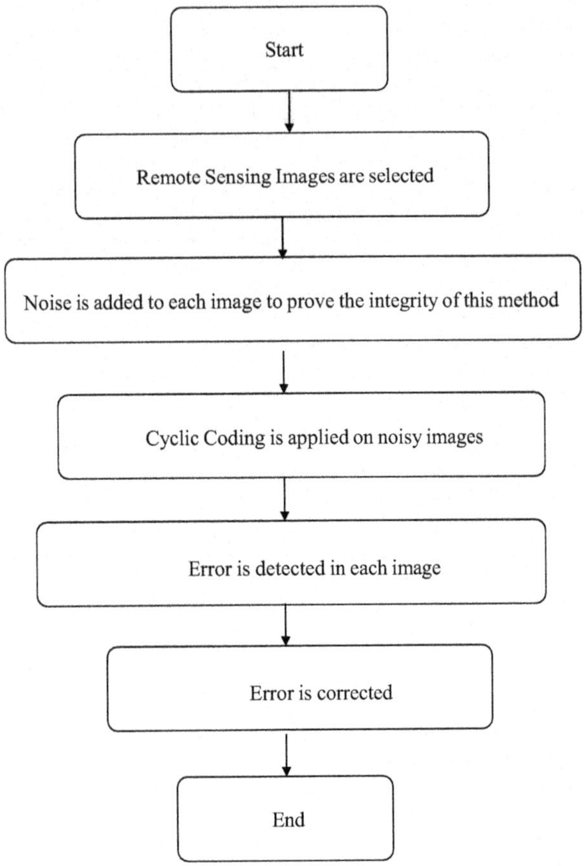

4 Result and Analysis

We have used six high-resolution remote sensing images for our research work (shown in Fig. 1) of pixel size 1024 × 1024. The entire work is done using MATLAB simulation software, 32 GB RAM, and Intel Core i5 Processor.

We have applied the down quantization process (8 bit to 4 bit) for simplicity.

To prove the effectiveness of our proposed method, we have added noise to the images. The noisy images are shown in Fig. 2.

Recovered images are shown in Fig. 3.

Fig. 1 Selected images

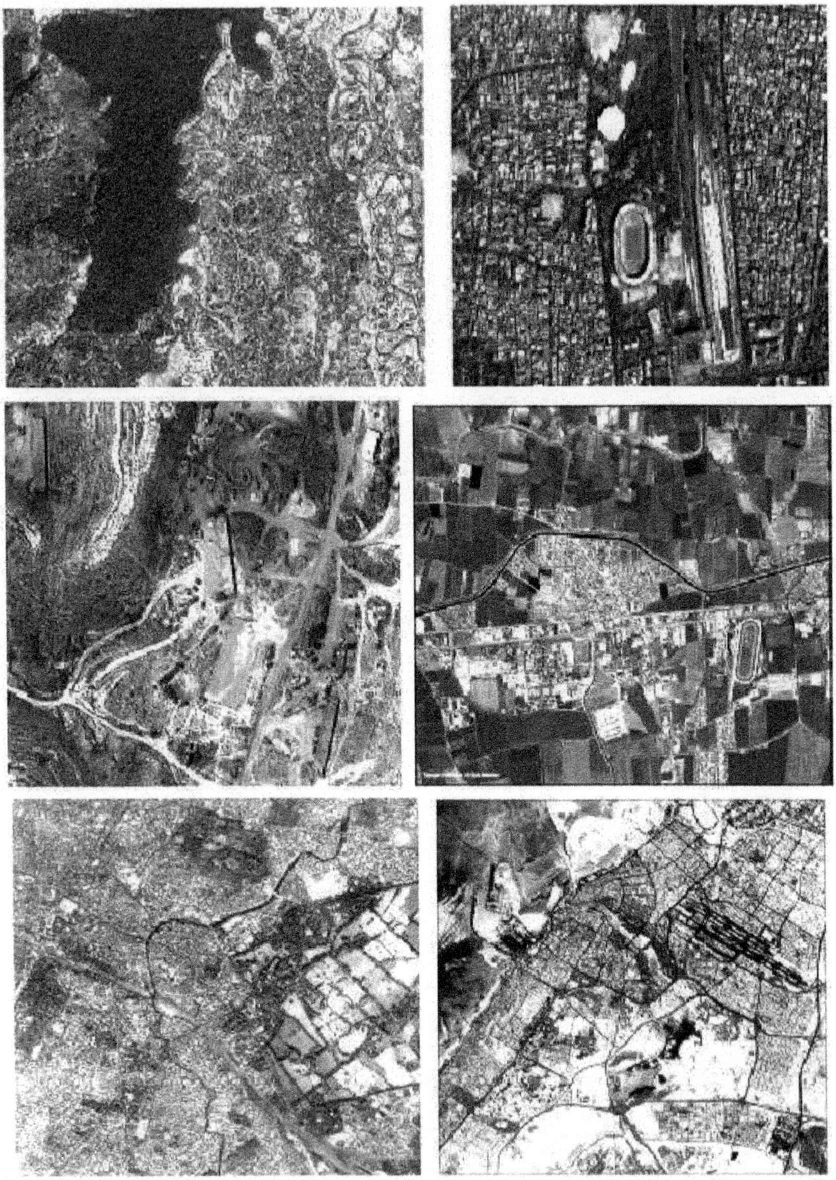

Fig. 2 Images with error

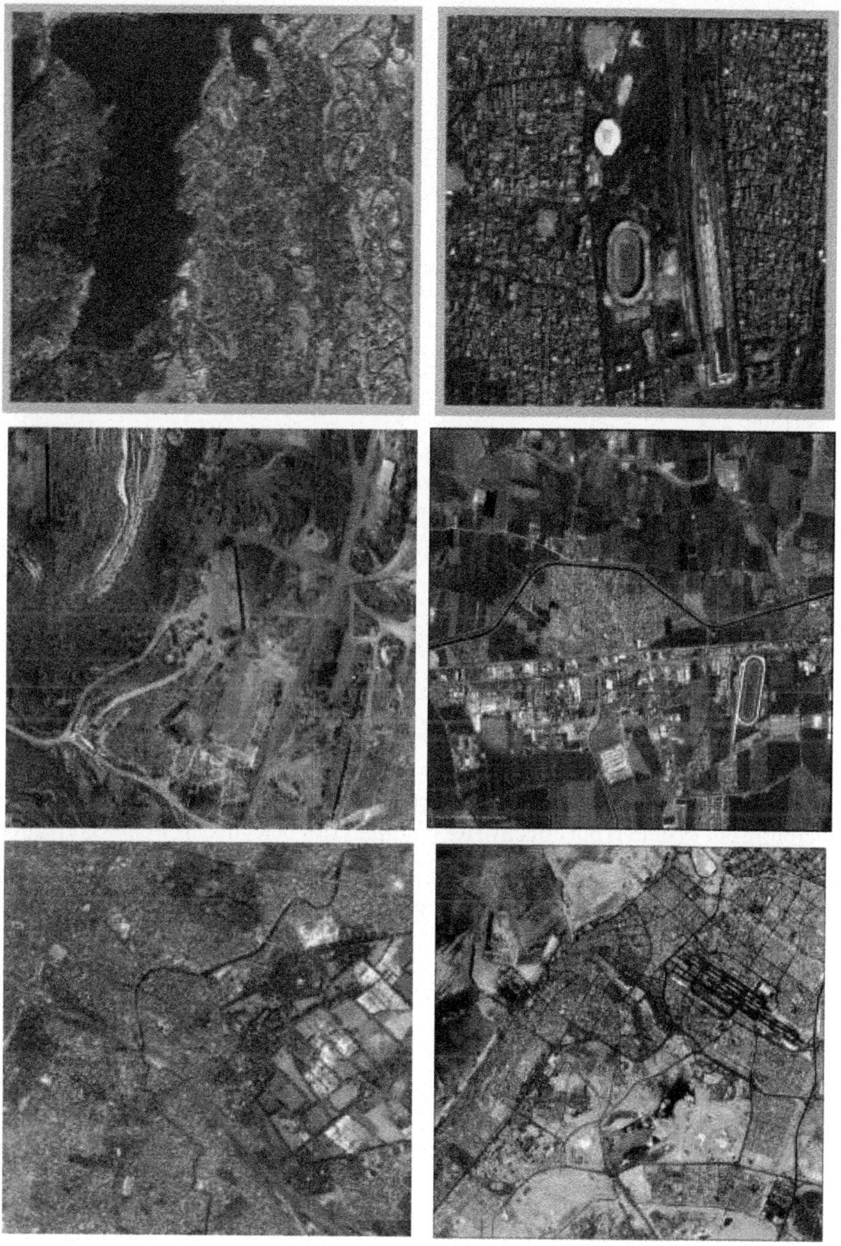

Fig. 3 Corrected images

Table 2 PSNR table

Image no.	PSNR
1	31.7
2	31.4
3	32.1
4	31.6
5	31.1
6	31.5

5 Advantage

See Table 2.

In previous works, cyclic coding was used in binary images and low-resolution grayscale images [14–17]. Application of cyclic coding in high-resolution images is a challenging task, In this research work, we have applied cyclic coding in high-resolution images and proved that cyclic coding can be effectively used in high-resolution images also.

6 Limitation

Though this system is effective for error detection as well as a correction there is one limitation. As cyclic code can detect only one error per message (here pixel), therefore our system is not effective for multiple error detection as well as correction.

7 Conclusion

In this communication, we have proposed an error detection and correction method for high-resolution remote sensing images using the systematic cyclic code. This system is really effective for the long-distance transmission of remote sensing images. If the images are affected by noise, this system can be eliminated. For simplicity, we have down quantized the selected images. PSNR value is greater than 31 which implied the satisfactory result of our method. Earlier research works are based on binary or low-resolution images whereas our system shows its effectiveness in high-resolution images. The only drawback of this system is that it is suitable to correct a single error. However, with the help of simulation software, we have proved the effectiveness of our proposed method.

References

1. B. Bose, D. Dey, A. Sengupta, N. Mulchandani, A. Patra, A novel medical image encryption using cyclic coding in Covid-19 pandemic situation. J. Phys.: Conf. Ser. 1797 (2021)
2. A. Patra, A. Saha, K. Bhattacharya, Multiplexing and encryption of images using phase grating and random phase mask. Opt. Eng. **59**(3), 033105–033116 (2020)
3. A. Patra, A. Saha, K. Bhattacharya, High-resolution image multiplexing using amplitude grating for remote sensing applications. Opt. Eng. **60**(7), 073104–073115 (2021)
4. A. Patra, A. Saha, K. Bhattacharya, Compression and multiplexing of medical images using optical image processing. Comput. Intell. Appl. Healthc. 63–71 (2020)
5. A. Patra, A. Saha, K. Bhattacharya, D. Chakraborty, Compression of high-resolution satellite images using optical image processing, in *Satellite Systems-Design, Modeling, Simulation and Analysis* (2021)
6. I. Öztürk, I. Sogukpınar, Analysis and comparison of image encryption algorithms. Trans. Eng. Comput. Tech. **3**, 1305–1313 (2004)
7. Y. Shuangyuan, L. Zhengding, H. Shuihua, Anasymmetric image encryption based on matrix transformation, in *IEEE International Symposium on Communications and Information Technology, ISCIT*, vol. 1 (2004), pp. 66–69
8. Y. Alsultanny, Image encryption by cipher feedback mode. Innov. Comput. Inf. Control. **3**, 589–596 (2007)
9. S. Kandar, A. Maiti, k-n secret sharing visual cryptography scheme for color image using random number. IJEST, 1851–58 (2011)
10. A. Nag, J. Singh, S. Khan, S. Biswas, D. Sarkar, P. Sarkar, Image encryption using affine transform and XOR operation. ICSCCN, 309–312 (2011)
11. M. Sokouti, B. Sokouti, S. Pashazadeh, M.-R. Feizi-Derakhshi, S. Haghipour, Genetic-based random keygenerator (GRKG): a new method for generating more-random keys for one-time pad cryptosystem. Neural Comput. Appl. **22**(7–8), 1667–1675 (2013)
12. M. Younes, A. Janatan, Animage encryption approach using a combination of permutation technique followed by encryption. IJCSNS **8**(4), 191–197 (2008)
13. S.H. Kamali, R. Shankeria, M. Hedayati, M. Rahmani, New modified versionof advance encryption standard based algorithm for image encryption, in *International Conference on Electronics and Information Engineering (ICEIE)* (2010), pp. v1-141–v1-5
14. J. Borges, S.T. Dougherty, C. Fernandez-Córdoba, R. Ten-Valls, Binary Images of Z_2 Z_4-additive cyclic codes. IEEE Trans. Inform. Theory **64**(12) (2018)
15. J. Wolfmann, Binary images of cyclic codes over Z_4. IEEE Trans. Inform. Theory **47**(5) (2001)
16. A. Rodrigues, A. Bhise, Reversible image steganography using cyclic codes and dynamic cover pixel selection, in *IEEE International Conference on Wireless Communications, Signal Processing and Networking (WiSPNET)* (2018)

17. G. Vega, Nonlinear cyclic codes over Z4 whose Nechaev-gray images are binary linear cyclic codes (2003), SSRN: https://ssrn.com/abstract=3158265
18. A. Patra, M. Das, A. Ghosal, A. Ghosh, I. Kushary, S. Roy, D. Chakraborty, Remote sensing image encryption and error detection using hamming code. J. Phys.: Conf. Ser. **2286**(012018) (2022)
19. A. Patra, A. Saha, K. Bhattacharya, Efficient storage and encryption of 32-Slice CT scan images using phase grating. Arab. J. Sci. Eng. (2022)

Segmentation of Used Biodegradable and Non-biodegradable Products of Covid 19 Patients Using Thermal Image Processing

Anirban Patra, S. K. Sohel Ahmed, and Sonali Sarkar

1 Introduction

For a long time, researchers have proposed a lot of techniques for segmentation of biodegradable and non-biodegradable materials. H. Jouhara et al. highlighted a method based on biological and physicochemical systems for separation of municipal domestic wastes [1]. This method is cost effective but the proposed method is suitable for domestic products only. A U. Zaman discussed in their research work the importance of separation of different materials to minimize its adverse effect on the environment [2]. Other researchers applied different methods in materials like plastic, solid waste fuel, crop, industrial waste, etc. [5–8]. In these methods, mostly chemical-based technology is used for isolation of different products. These processes are time-consuming for detection. Moreover, the chemical activities have some side effects too. The use of image processing is widely used for the last few decades to separate different types of materials based on their characteristics. The K Means algorithm is a conventional technique for segmentation [9–12].

X. Zheng et al. utilized the adaptive K Means algorithm to get accurate results. N. Akhtar utilized K Means clustering along with neurotrophy in their research work. Histogram processing, edge detection, and thresholding are also effective tools for image segmentation [13, 14]. K. Qin and K. Xu utilized cloud models in histogram processing for separating different parts of images. Spatial metrics are used for analyzing urban land use which is an important part of land cover analysis [15]. Multiscale geoscience segmentation is an effective tool for extracting urban functional zones from VHR Satellite Images [16].

A. Patra (✉) · S. K. Sohel Ahmed
Department of ECE, JIS College of Engineering, Kalyani, West Bengal, India
e-mail: anitublu@gmail.com

S. Sarkar
Department of Basic Science, Swami Vivekananda Institute of Science and Technology, Kolkata, India

A. Saha et al. (eds.), *Optical to Terahertz Engineering*, Progress in Optical Science and Photonics 23, https://doi.org/10.1007/978-981-99-0228-6_11

The use of neural networks in addition to conventional image segmentation techniques in satellite and aerial images has been proposed by few researchers. The rapid use of artificial intelligence enhanced the segmentation purpose nowadays. The application of convolution neural network in land cover classification is described by K. Nogueira et al. [17]. A. Patra et al. highlighted the use of different optical information processing methods for the same purpose [18–22]. Sinusoidal amplitude and phase grating along with phase mask is used. Moreover, they have utilized the alpha blending process (pixel value modification) for watermarking segmented images (medical, satellite images).

Rubin observed that, although color is one of the most interesting and integral parts of vision, most models and methods of colorimetric available to describe and quantify color have been developed outside of optometry [23]. Apriyanti et al. developed a method automated color detection in orchids using color labels and deep learning [24]. Color image process using principal element analysis describes that the color recognition involves comparison of every pixel within the metric and leads to the dominant color because the color of the given object [25].

Shriram et al. proposed new real time color recognition features, i.e., extracting primary colors for the aim of vision-based human–computer interaction. Vision-based human–computer interaction can be achieved by analyzing segmental primary color regions primarily focused on color-based image segmentation and vision primarily based color recognition by addressing these difficulties [26]. Xiong et al. proposed a novel real-time color image segmentation method is, which is based on color similarity in RGB color space [27]. Tsai and Tseng proposed color detection method based on hue, saturation, and lightness (HSL) color space [28].

Most of the above-mentioned systems used sensor-based systems for segmentation. The longevity of the sensor is low. Hence, these systems do not cost much as well as are time-effective. In this communication, we have applied thermal imaging by selecting a thresholding value for the classification of the bio- and non-biodegradable materials. In digital image processing, pixel intensity value takes a major role in the segmentation of images. The entire system is based on pixel intensity value-oriented segmentation.

In the first stage, the selected images are converted into thermal images by special thermal processing application software. Following, the pixel intensity difference is calculated from the pixel value of a reference point [0,0] (this reference value is selected as it is the corner coordinate). As each color of the thermal images lies within a specified range. Therefore, by this method, we have got different values for different colors. Thereafter based on the result, we have primarily classified the results into two categories: bio- and non-biodegradable items.

2 Biodegradable and Non-biodegradable Materials

In the present day, the advancement of Science and Technology has improved our quality of life and has eased our lifestyles. But human beings for their own benefit have exploited the environment as well as the lives of other living beings of the earth.

Human beings have ruthlessly cut down trees, forest, polluted the air with exhaust gases of industries and vehicles, polluted the water with poisonous chemicals and the ground or soil with excess fertilizers, plastics, heavy metals etc. This has resulted in the damage or decrease in the habitat of other living organisms and degradation of the environment which not only poses danger to other organisms but to human beings as well.

We have not done sustainable development earlier, i.e. judicious use of the natural resources and optimum development of our lives which would have harmed the environment less. This has resulted in our environment to become more polluted for our next generation.

Some waste materials like plastic, bottles may be convenient for daily use but these waste materials after their disposal in the environment cause pollution of air, water and land.

Among these waste materials some are dangerous while others are not. This is because such waste materials can be largely categorized into biodegradable waste materials and non-biodegradable waste materials.

2.1 *Biodegradable*

Biodegradable materials are those natural materials which can be broken down and degraded through natural processes by bacteria and other soil microorganisms and get absorbed into the environment and do not cause any major damage or degradation of the environment.

2.2 *Non-biodegradable*

Non-biodegradable materials are man-made materials which cannot be broken down by natural processes by microorganisms and do not decay in the environment.

They remain in the environment for infinite time and turn into a huge heap of garbage.

They remain in the environment, degrade it and pose danger to living organisms and the environment.

Exp: Plastic, Polyethene, bottles, concrete, chemicals etc.

Still the biodegradable substances are called biodegrade wastes as though they are less riskier for the environment, but eventually the dumping of these huge wastes cause the release of obnoxious gases and some pollute the air due to incineration of these wastes.

So these biodegradable wastes are either converted to compost which may be used as manure or biogas may be obtained from them or simply dumped into landfills away from city.

Non-biodegradable wastes do not decay through natural process. So to decrease the waste we have started practicing the 3Rs, Reduce (reduce the waste), Refuse (Refuse use of plastics etc.) and Recycle (Recycle some materials like glass, metals, polytene etc. instead of disposing them directly).

3 Long Term Effect of Waste Materials on the Environment

The world during its advancement with the help of technology, resulted in releasing and dumping of huge quantity of waste materials. Wastes categorized as municipal, industrial, bio-medical and hazardous wastes cause multi-dimensional harm to the environment. They cause loss of useful materials, scarcity of resources, utilization of huge amount of energy, high cost of collection, treatment and disposal of the wastes.

These wastes also generate huge quantity of greenhouse gases like carbon dioxide and methane that increase the temperature of the earth and cause global warming. This has a significant effect on the overall heat of the earth surface that are causing climatic changes throughout the world including melting of the ice of the arctic region etc.

Hence the importance of identification, disposal, treatment and elimination of these wastes in the environment is to ensure sustainable development of the environment.

4 Methodology

Let us assume the selected image is represented as $g_x(x, y)$. A pixel intensity value in this selected image is represented by $j(x, y)$.

Let the intensity of the pixel at coordinate (0, 0) [which we have selected as reference] is $j(0, 0)$.

The difference between each pixel from the reference is

$$(x, y) = (x, y) - (0, 0) \tag{1}$$

Based on the average value of (x, y), we have classified it into two zones $e1$ and $e2$ respectively.

Let us assume that, threshold value $= th$

If $e(x, y) > th$

$$(x, y) = e1 \tag{2}$$

If $e(x, y) < th$

$$e(x, y) = e2. \tag{3}$$

The entire methodology is explained with the help of a flow chart which is shown in Fig. 1.

From our selected images, we have applied a threshold-based image segmentation process. In threshold-based, we have set a reference value for comparison.

We have primarily calculated it for all two materials. The more specific division is also possible by varying the threshold values on the selected thermal images.

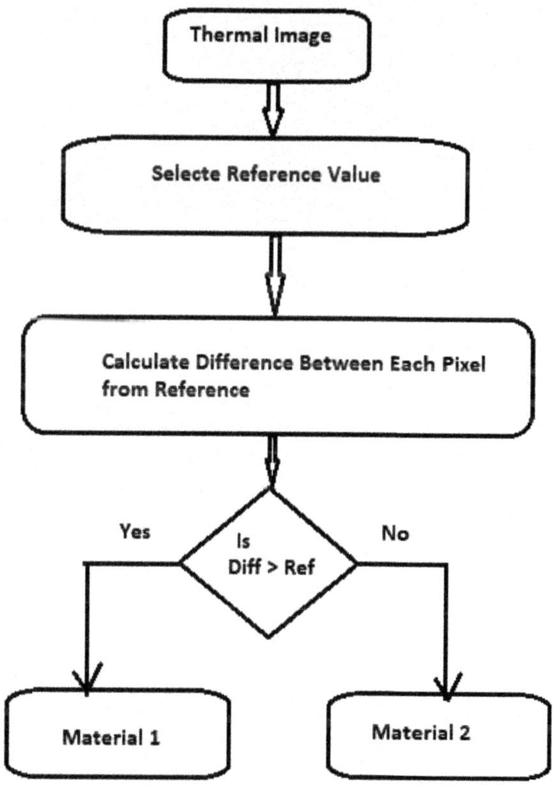

Fig. 1 General flow chart

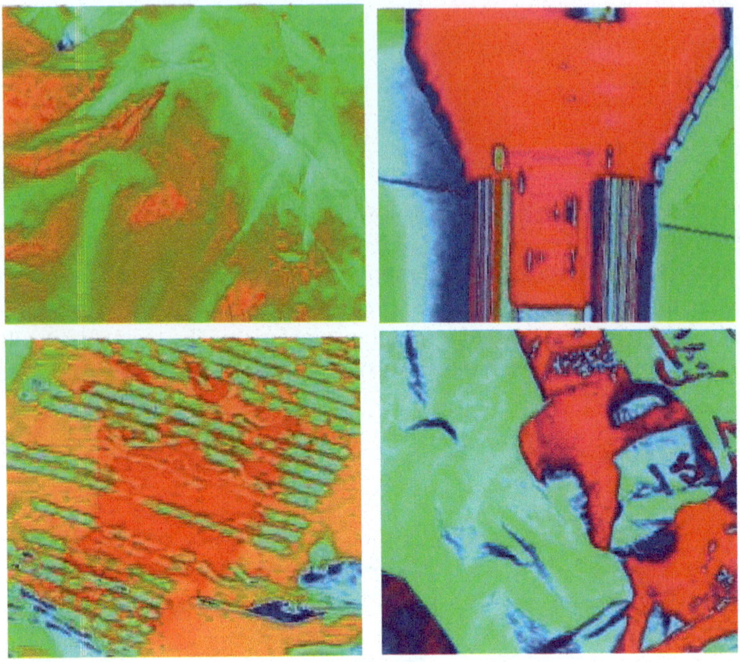

Fig. 2 Non-biodegradable materials

5 Result

In our research work, we have worked with a few biodegradable and non-biodegradable materials. The pixel resolution of our selected images is 512 × 512. The output of the thermal images of biodegradable and non-biodegradable materials is displayed in Figs. 2 and 3.

6 Analysis

Initially, we categorized the materials as biodegradable and non-biodegradable materials and captured thermal images of them. Selected non-biodegradable materials are mainly plastics and selected biodegradable materials are vegetables and paper. According to our methodology, we have calculated the intensity difference of pixel intensity values of each coordinate of all images from the reference value (coordinate 0,0) of the image. The result of the pixel intensity difference is shown in Table 1.

From the table, it is clear that if we set the threshold value as 50, we can easily separate bio and non-biodegradable materials. We have performed our research work

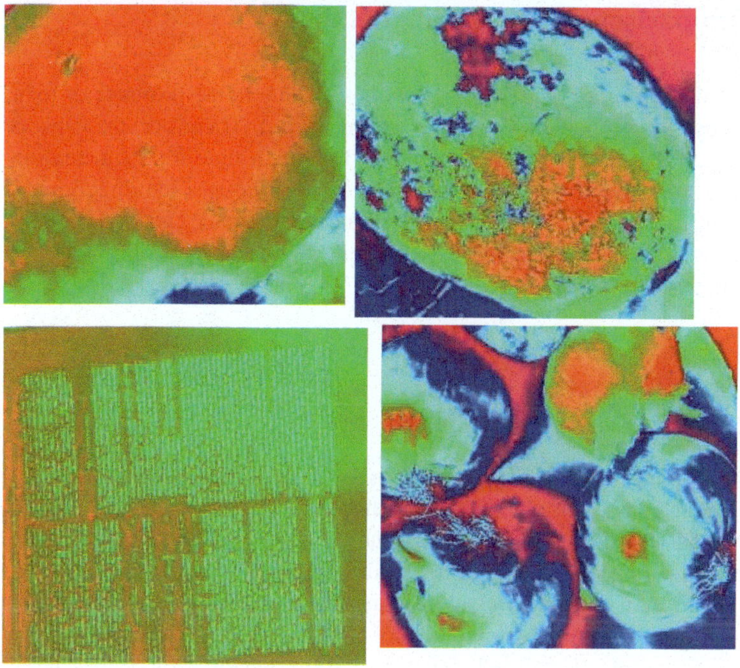

Fig. 3 Bio-degradable materials

Table 1 Comparison table

Sl No	Image type	Intensity difference
1	Non-biodegradable	39
2	Non-biodegradable	47
3	Non-biodegradable	36
4	Non-biodegradable	33
5	Biodegrable	68
6	Biodegrable	79
7	Biodegrable	61
8	Biodegrable	57

in MATLAB 2014b software. For thermal imaging, we have used thermal image applications.

7 Comparison with Existing Methods

Most of the existing systems are sensor-based. Due to the presence of toxic materials, sensors are unable to work for a long period. Therefore, the sensors are to be changed frequently. Moreover, it takes too much time for segmentation. So, it is neither cost-effective nor time effective. It also violates the direction of WHO, as human intervention is required to operate the process. Our system works on a thermal imaging camera which lasts for many years. In addition, it takes less time for segmentation (a few seconds).

8 Conclusion

In a post-pandemic situation, it is essential to separate used bio and non-biodegradable materials of Covid 19 patients on a non-contact basis. In this situation, our method would be helpful for this purpose. No human intervention is required here as it separated both types of materials with the help of thermal imaging. In our method, we have calculated pixel intensity difference from the (0,0) position but it can be measured from any coordinate of the images. In future communication, we will use more sophisticated techniques. By varying the threshold values, it is possible to sub the grouping of the products too. Our proposed technique is simple as it has the potential of separating a lot of materials within a few seconds. Hence it is a time as well as cost-saving process. Using machine learning and artificial intelligence, a modified system of this technology is possible which will be discussed in future communication.

References

1. H. Jouhara, D. Czajczynska, H. Ghazal, R. Krzyzy, L. Anguilano, A.J. Reynolds, N. Spencer, Municipal waste management systems for domestic use. Energy **139**, 485–506 (2017)
2. U. Zaman, A comprehensive study of the environmental and economic benefits of resource recovery from global waste management systems. J. Clean Prod. **124**, 41–50 (2016)
3. T. Karak, R.M. Bhagat, P. Bhattacharyya, Municipal solid waste generation, composition, and management: the world scenario. Crit. Rev. Environ. Sci. Technol. **42**, 1509–1630 (2012)
4. H. Zhou, A. Meng, Y. Long, Q. Li, Y. Zhang, An overview of characteristics of municipal solid waste fuel in China: physical, chemical composition and heating value. Renew. Sustain. Energy Rev. **36**, 107–122 (2012)
5. R. Couth, C. Trois, Carbon emissions reduction strategies in Africa from improved waste management: a review. Waste Manag. **30**, 2336–2346 (2010)
6. G. Davis, J.H. Song, Biodegradable packaging based on raw materials from crops and their impact on waste management. Ind. Crops Prod. **23**, 147–161 (2006)
7. G. Kale, T. Kijchavengkul, R. Auras, M. Rubino, S.E. Selke, S.P. Singh, Composability of bioplastic packaging materials: an overview. Macromol. Biosci. **7**, 255–277 (2012)
8. M. Vaverková, D. Adamcová, J. Zloch, How do degradable/biodegradable plastic materials decompose in home composting environment. J. Ecol. Eng. **15**, 82–89 (2012)

9. N. Dhanachandra, K. Manglem, Y.J. Chanu, Image segmentation using K-means clustering algorithm and subtractive clustering algorithm. Procedia Comput. Sci. **54**, 764–771 (2012)
10. X. Zheng, Q. Lei, R. Yao, Y. Gong, Q. Yin, Image segmentation based on adaptive *K*-means algorithm. EURASIP Journal on Image and Video Processing , volume 68; 2018
11. D.J. Bora, A.K. Gupta, A novel approach towards clustering based image segmentation. IJESE **2**(Issue-11), (2014). ISSN: 2319–6378
12. N. Akhtar, N. Agarwal, A. Burjwal, K-mean algorithm for image segmentation using neutrosophy, in *2014 International Conference on Advances in Computing, Communications and Informatics* (2014)
13. G.M. Hadi, N.H. Salman, Medical image segmentation based on image histogram and clustering technique. Int. J. Graph. Image Process. **5**(Issue 1), (2015)
14. K. Qin, K. Xu, Image segmentation based on histogram analysis utilizing the cloud model. Comput. Math. Appl. **62**(7), 2824–2833 (2011)
15. M. Herald, U. Clark, Spatial metrics and image texture for mapping urban land use. Geogr. Photogram. Eng. Remote Sens. (2003)
16. X. Zhang, S. Du, Q. Wang, W. Zhou, Multiscale geoscene segmentation for extracting urban functional zones from VHR satellite images. Remote Sens. **10**, 281 (2018)
17. K. Nogueira, O.A. Penatti, D. Santos, Towards better exploiting convolutional neural networks for remote sensing scene classification. Pattern Recogn. **61**, 539–556 (2017)
18. A. Patra, K. Anubala, M. Ghosh, D. Chakraborty, Study on land cover analysis of using satellite and arialimages. Int. J. Sci. Res. Eng. Dev. **3**, 1081–1084 (2020)
19. A. Patra, A. Saha, K. Bhattacharya, Multiplexing and encryption of images using phase grating and random phase mask. Opt. Eng. **59**, 033105 (2020)
20. A. Patra,A. Saha, A.K. Chakraborty, A simple approach to watermarking of multiple grayscale images using alpha blending. IRJET **04**(Issue 03), (2017)
21. A. Patra, A. Saha, K. Bhattacharya, Efficient storage and encryption of 32-slice CT scan Images using phase grating. Arab. J. Sci. Eng. (2022)
22. A. Patra, B. Bose, A. Sengupta, D. Dey, Effective storage and transmission of satellite and medical images using optical information processing. IJIREEICE **10**(4), (2022)
23. Hasting, G., Rubin, A., Color spaces—a review of historic and modern color models. Afr. Vision Eye Health **71** (2012). https://doi.org/10.4102/aveh.v71i3.76
24. D.H. Apriyanti, L.J. Spreeuwers, P.J.F. Lucas, R.N.J. Veldhuis, Automated color detection in orchids using color labels and deep learning. PLoS One **16**(10), (2021)
25. A. Abadpour, S. Kasaei, Principal color and its application to color image segmentation. Scientia Iranica **15**, 238–245 (2008)
26. D. Senthamaraikannan, S. Shriram, J. William, Real time color recognition. Int. J. Innov. Res. Electr. Electron. Instrum. Control Eng. **2**(3), (2014)
27. N. Xiong, Y. Shen, K. Yang et al., Color sensors and their applications based on real-time color image segmentation for cyber physical systems. J. Image Video Proc. **2018**, 23 (2018)
28. S.-H. Tsai, Y.-H. Tseng, A novel color detection method based on HSL color space for robotic soccer competition. Comput. Math. Appl. **64**(5), 1291–1300 (2012)

The manufacturer's authorised representative in the EU is Springer Nature Customer Service Centre GmbH, Europaplatz 3, 69115 Heidelberg, Germany. If you have any concerns regarding our products, please contact ProductSafety@springernature.com

Printed and bound by CPI Group (UK) Ltd, Croydon, CR0 4YY
30/09/2026
02225308-0017